SRA
Connecting Math Concepts

Level B Teacher's Guide

COMPREHENSIVE EDITION

A DIRECT INSTRUCTION PROGRAM

Mc Graw Hill **Education**

Bothell, WA • Chicago, IL • Columbus, OH • New York, NY

Acknowledgments

The authors are grateful to the following people for their input in the field-testing and preparation of *SRA Connecting Math Concepts: Comprehensive Edition Level B:*

Amilcar Cifuentes
Bob Dixon
Crystall Hall
Joanna Jachowicz
Dan Johnston
Debbie Kleppen
Margie Mayo
Michael McGlauhlin
Alicia Smith
Jason Yanok

MHEonline.com

Send all inquiries to:
McGraw-Hill Education
4400 Easton Commons
Columbus, OH 43219

ISBN: 978-0-02-103593-9
MHID: 0-02-103593-8

Printed in the United States of America.

5 6 7 8 9 QVS 18 17 16 15

*The **McGraw·Hill** Companies*

Contents

Program Overview

The *Connecting Math Concepts: Comprehensive Edition Series*

Connecting Math Concepts: Comprehensive Edition is a six-level series that will accelerate the math learning performance of students in grades K through 5. Levels A through F are suitable for regular-education students in Kindergarten through fifth grade. The series is also highly effective with at-risk students in any grade.

Connecting Math Concepts: Comprehensive Edition is based on the fact that understanding mathematics requires making connections

- Among related topics in mathematics, and
- Between procedures and knowledge.

Connecting Math Concepts: Comprehensive Edition does more than expose students to connections. It stresses understanding and introduces concepts carefully, then weaves them together throughout the program. Once something is introduced, it never goes away. It often becomes a component part of an operation that has several steps.

The organization of *Connecting Math Concepts: Comprehensive Edition* is powerful because lessons have been designed to:

1. Teach explicit strategies that all students can learn and apply.

2. Introduce concepts at a reasonable rate, so all students make steady progress.

3. Help students make connections between important concepts and key ideas.

4. Provide the practice needed to achieve mastery and understanding.

5. Meet the math standards specified in the Common Core State Standards for Mathematics.

The program's Direct Instruction design permits significant acceleration of students' performance for both high performers and students who are at risk. The instructional sequences are the same for all students, but the rate at which students proceed through each level should be adjusted according to students' performance. Higher performers proceed through the levels faster. Lower performers receive more practice. Benchmark in-program Mastery Tests provide information about how well students are mastering what has been taught most recently. Students' daily performance and test performance disclose whether they need more practice or whether they are mastering the material on the current schedule of lesson introduction.

The program enables the teacher to teach students at a faster rate and with greater understanding than they probably ever achieved before. The scripted lessons have been shaped through extensive field-testing and classroom observation. The teacher individualizes instruction to accommodate different groups that make different mistakes and require different amounts of practice to learn the material.

Introduction to 2012 *CMC Level B*

CMC Level B is designed for students who have successfully completed *CMC Level A* or who pass the Placement Test for *CMC Level B* (See page 100.).

CMC Level B instruction meets all requirements of the first-grade Common Core State Standards for Mathematics.

Program Information

The following summary table lists facts about 2012 *CMC Level B.*

Students who are appropriately placed in *Level B*	Pass Placement Test (p. 100)
How students are grouped	Instructional groups should be as homogeneous as possible.
Number of lessons	• 125 regular lessons • 12 Mastery Tests • 2 Cumulative Tests
Schedule	• 40 minutes for structured work • Additional 10 minutes for students' Independent Work • 5 periods per week

Teacher Material	• Teacher's Guide
	• Presentation Book 1:
	    Lessons 1–40, Tests 1–4, Answer Keys 1–40
	• Presentation Book 2:
	    Lessons 41–80, Tests 5–8, Answer Keys 41–80, Cumulative Test 1 (Lessons 1–60)
	• Presentation Book 3:
	    Lessons 81–125, Tests 9–12, Answer Keys 81–125, Cumulative Test 2 (Lessons 1–125)
	• Board Displays CD

Student Material	• Workbook 1: Lessons 1–60 • Workbook 2: Lessons 61–125 • Student Assessment Book: Tests 1–12, Remedies worksheets for Mastery Tests 1–12, Cumulative Tests 1 and 2
In-Program Tests	12 ten-lesson Mastery Tests • Administration and Remedies are specified in the Teacher Presentation Books. • Tests and Remedies worksheets are in the Student Assessment Book.
Optional Cumulative Tests	2 Cumulative Tests • Administration is specified in Teacher Presentation Books 2 and 3. • Tests are in the Student Assessment Book.
Additional Teacher/ Student Material	• Student Practice Software (accessed via ConnectED) • Math Fact Worksheets (Online Blackline Masters via ConnectED) • Access to *CMC* content online via ConnectED • *SRA 2Inform* available on ConnectED for online progress monitoring

TEACHER MATERIAL

The teacher material consists of:

1. **The Teacher's Guide:** This guide explains the program and how to teach it properly. The Scope and Sequence chart on pages 8–11 shows the various tracks (topics or strands) that are taught; indicates the starting lesson for each track/strand; and shows the lesson range. This guide calls attention to potential problems and provides information about how to present exercises and how to correct specific mistakes the students may make. The guide is designed to be used to help you teach more effectively.

2. **Three Teacher Presentation Books:** These books specify each exercise in the lessons and tests to be presented to the students. The exercises provide scripts that indicate what you are to say, what you are to do, the responses students are to make, and correction procedures for common errors. (See Teaching Effectively, **Using the Teacher Presentation Scripts,** for details about using the scripts.)

3. **Answer Key:** The answers to all of the problems, activities, and tests appear in the Answer Key in the back of each TPB to assist you in checking the students' classwork and Independent Work and for marking tests. The Answer Key also specifies the remedy exercises for each test and provides a group summary of test performance.

4. **Board Displays CD:** The teacher materials include a Board Displays CD, which shows all the displays you present during the lessons. This component is flexible and can be utilized in different ways to support the instruction—via a computer hooked up to a projector, to a television, or to any interactive white board. The electronic Board Displays are available online on ConnectED or on CD. You can navigate through the displays with a touch of the finger if you have an interactive white board, or navigate using a mouse (wired or wireless), or a remote control.

5. **Practice Software:** The *CMC Level B* Practice Software provides students additional practice with the skills and concepts taught in *CMC Level B*. It is a core component for meeting several Common Core State Standards for Mathematics. Students apply their skills to tasks presented onscreen. The tasks are governed by an algorithm that adjusts the amount of practice students receive according to how well they perform. Games and reward screens provide students with reinforcement for meeting performance goals. The software is organized into blocks, each presenting activities for a twenty-lesson segment of the program as students proceed through the lessons.

   The Math Facts strand of the software is organized into sets of facts that follow the instructional sequence in the lessons. It is designed to facilitate continuous review and reinforcement of the math facts as they are introduced and practiced. It is available via ConnectED with 10 student seat licenses per every teacher materials kit purchase.

6. **ConnectED:** On McGraw-Hill/SEG's ConnectED platform you can plan and review *CMC* lessons and see correlations to Common Core State Standards for Mathematics. Access the following *CMC* materials from anywhere you have an Internet connection: PDFs of the Presentation Books, an online planner, online printable versions of the Board Displays CD, student Practice Software (online version requires separately purchased student licenses), Math Fact Worksheets BLMs, eBooks of the Teacher's Guides, and correlations. *CMC* on ConnectED also features a progress monitoring application called *SRA 2Inform* that stores students' data and provides useful reports and graphs about students' progress. Refer to the card you received with your teacher materials kit for more information about redeeming your access code, good for one six-year teacher subscription and 10 student seat licenses, which provide access to the Practice Software and eTextbook.

STUDENT MATERIAL

The student materials include a set of two Workbooks for each student, and a Student Assessment Book. The Workbooks contain writing activities, which the students do as part of the structured presentation of a lesson and as independent seatwork. The Student Assessment Book contains material for the Mastery Tests as well as test Remedies worksheet pages and optional Cumulative Test pages.

   Workbook 1: Lessons 1–60

   Workbook 2: Lessons 61–125

   Student Assessment Book: Mastery Tests 1–12, Cumulative Test 1 and 2, and test Remedies worksheets for Mastery Tests

WHAT'S NEW IN 2012 *CMC LEVEL B*

Most instructional strategies are the same as those of the earlier *CMC* editions; however, the

procedures for teaching these strategies have been greatly modified to address problems teachers had teaching the content of the previous editions to at-risk students. The 2012 edition of *CMC Level B* has also been revised on the basis of field-testing.

1. The 2012 edition provides far more oral work than earlier editions. This work is presented as "hot series" of tasks. The series are designed so that students respond to ten or more related questions or directions per minute; therefore, these series present a great deal of information about an operation or discrimination in a short period of time.

2. The content is revised so that students learn not only the basics but also the higher-order concepts. The result is that first-grade students who complete *CMC Level B* are able to work the full range of problems and applications that define understanding of first-grade math.

3. The hallmark of Direct Instruction mathematics programs is that they teach all the component skills and operations required to provide a solid foundation in topics involving place value and operations, money, geometry, measurement and data, and word problems. The *CMC Level B* program addresses all standards specified in the Common Core State Standards for Mathematics for first grade. (pages 10–11 and 94–97.)

4. *CMC Level B* offers support/enhancements, including technology components, for teachers and students. These enhancements include displays in the Teacher Presentation Books, a Board Displays CD (also available online), Workbook and Answer Key pages reduced in the Teacher Presentation Book, a Student Assessment Book with all program assessments in one location, *SRA 2Inform* for online progress monitoring, Student Practice Software, and the ability to plan and review lessons online via ConnectED.

The Structure of Connecting Math Concepts Level B

Connecting Math Concepts Level B is appropriate for students who complete *Connecting Math Concepts Level A* or who pass the *Connecting Math Concepts Level B* Placement Test.

Connecting Math Concepts Level B has two starting points: Lesson 1 and Lesson 16. Lessons 1–15 are designed to acquaint new students with the conventions that continuing students learned in *Level A*. Continuing students start at Lesson 16. Lessons are designed to review facts and other information presented in *Level A* and to introduce new material at a rate that would be appropriate for both the continuing students and the students who started at Lesson 1.

If you have a group that has new and continuing students, the simplest solution is to start all students at Lesson 1 and proceed as quickly as the lower performers, who are appropriately placed in the program, are able to proceed.

Reproducible copies of the Placement Test appear on pages 105–106 of this guide.

SCHEDULING

The program contains 125 lessons and 12 in-program Mastery Tests. The ideal goal is to teach one lesson each period. If students are not firm on content that is being introduced, you will need to repeat parts of lessons or entire lessons. Particularly early in the program, you will need to repeat entire lessons because students will perform much better on subsequent lessons if all lessons are taught to mastery.

Also, some lessons are longer and may require more than a period to complete. Following long lessons, try to get back on a schedule of teaching a lesson a day. This pattern assures that students receive daily practice in skills or operations that have been recently introduced.

The program is to be taught daily. Periods for structured work are 40 minutes. Students need an additional 10 minutes or more to complete their Independent Work. If the Independent Work cannot be completed in school, it may be assigned as homework, but this is not an attractive alternative, particularly on Fridays. It's important for students to bring back their work on the following school day.

A final note: *CMC Level B* is not designed as a supplemental program and should not be used as one.

HOW THE PROGRAM IS DIFFERENT

Connecting Math Concepts Level B differs from traditional approaches in the following ways:

Field Tested

CMC Level B has been shaped through field testing and revision based on difficulties students and teachers encountered. The field-test philosophy is that if teachers or students have trouble with material presented, the program is at fault. Revisions are made to alleviate observed problems.

The field-test results of *Connecting Math Concepts Level B* disclose that if the teacher implements the program according to the presentation detail provided for each exercise, students will learn the content and become proficient in the content to advance to the next levels of math instruction.

Organization and Instructional Design

CMC Level B represents a sharp departure from the idiom of how to teach math through "discovery" or even through programs that have a progression of "units" and some form of cumulative review. These programs severely underestimate the amount of practice students need to attain fluency with problem-solving steps, such as translating word problems into equations, solving them, and answering the question the problem asks. The programs also don't have a good scheme for teaching the essential component skills of complex operations before these operations are introduced. All levels of *CMC* strictly follow the practice of first introducing all the component discriminations and skills students need, then combining them into an operation. All that remains to be taught is the sequence of steps and the difference between the newly taught operation and similar operations students have been taught.

The design of lessons is based largely on the following considerations:

a. During a period, it is not productive to work only on a single topic. If a lot of new information is being presented, students may become overwhelmed. A better procedure— one that has been demonstrated to be superior in studies of learning and memory—is to distribute the practice; so instead of working for 40 minutes on a single topic, students work each day for possibly 10 minutes on each of four or five topics.

b. When full-period topics are presented, it becomes very difficult for a teacher to provide sufficient practice and review on the latest skills that have been taught. A more productive organization works on a particular problem type for a small part of many consecutive lessons. This organization presents each topic as a **track.**

c. The lessons that result from the track organization rather than the single-topic organization are designed so that only about 10–15% of the lesson material is new—introduced for the first time. The rest is either work on problem types that have been introduced in the preceding lessons, slight expansions, or new applications that build on what was taught earlier. The **Tracks** section of this guide (page 32) presents an overview of the design strategies used in each track to minimize the amount of new material that is taught.

Scripted Presentations

All exercises in each lesson are scripted. The script indicates the wording you use in presenting the material and correcting students' errors. Once you are familiar with the program, you may deviate some from the exact wording; however, until you know why things are phrased as they are, you should follow the exact wording. The most common mistakes teachers make in presenting the material is to rephrase some instructions. Later, when the original instructions become components of more complicated operations, the students are not prepared to respond to steps that have variant wording.

In *Connecting Math Concepts Level B,* you first present material in a structured sequence that requires students to respond verbally. This technique permits you to present tasks at a high rate so it is very efficient for teaching. Also, it provides you with information about which students are responding correctly and which need more repetition.

Typically, after students respond to a series of verbal tasks, you present written work.

Here's an exercise with a verbal series in which students identify whether the big number or a small number is missing, and, therefore, whether they add or subtract to find the missing number.

b. (Display:) W   [37:3A]

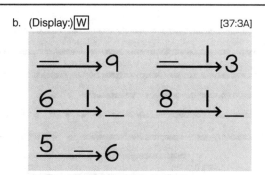

Some of these families have a missing small number. Others have a missing big number. Here's a rule: If a small number is missing, you minus.

- What do you do if a small number is missing? (Signal.) *Minus.*

c. (Point to ⎯¹→9.) Is a small number missing? (Signal.) *Yes.*

- So do you minus? (Signal.) *Yes.*

d. (Point to 6⎯¹→_.) Is a small number missing? (Signal.) *No.*

- So do you minus? (Signal.) *No.*
  So you plus.

- What do you do? (Signal.) *Plus.*

e. (Point to 5⎯→6.) Is a small number missing? (Signal.) *Yes.*

- So do you minus? (Signal.) *Yes.*

f. (Point to ⎯¹→3.) Is a small number missing? (Signal.) *Yes.*

- So do you minus? (Signal.) *Yes.*

g. (Point to 8⎯¹→_.) Is a small number missing? (Signal.) *No.*

- So do you minus? (Signal.) *No.*
- What do you do? (Signal.) *Plus.*

h. Now you're going to tell me the problem for each family.

- (Point to ⎯¹→9.) Is a small number missing? (Signal.) *Yes.*

- So do you minus? (Signal.) *Yes.*

- Start with the big number and say the problem. (Touch.) *9 minus 1.*

i. (Point to 6⎯¹→_.) Is a small number missing? (Signal.) *No.*

- So do you minus? (Signal.) *No.*

- Start with the small number and say the problem. (Touch.) *6 plus 1.*

j. (Point to 5⎯→6.) Is a small number missing? (Signal.) *Yes.*

- So do you minus? (Signal.) *Yes.*

- Start with the big number and say the problem. (Touch.) *6 minus 5.*

from Lesson 37, Exercise 3

A great advantage of sequences like these is that if students later make mistakes involving the discriminations the series addressed, you have information about exactly how to correct the mistakes. For instance, if students make mistakes of doing the wrong operation, your correction would be to ask:

- Is a small number missing?
- So what do you do if a small number is missing?

The scripted presentation is designed to help you present the key discriminations quickly and with consistent language, which helps maximize the efficiency of your teaching.

Language

Some conventions of *Connecting Math Concepts Level B* may initially give the impression that the presentations lack integrity because they do not always use the traditional language associated with the content. The reason is that what is being taught occurs in stages over many lessons, not all at once in a single lesson or several lessons. The language students need to solve traditional problems will ultimately be taught. The general format of introduction, however, calls for a minimum of vocabulary and a strong emphasis on demonstrating how the operation works, what the discriminations are, and which steps are needed to solve problems. Vocabulary that is not essential to solving a problem type will probably not be introduced.

Scope and Sequence for Connecting Math Concepts Level B

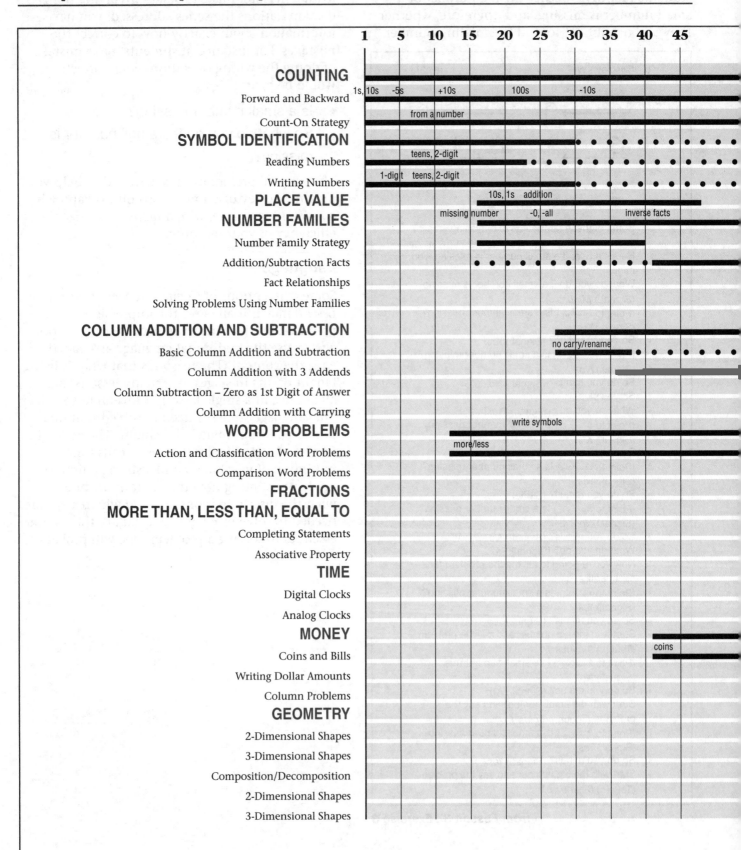

	1	5	10	15	20	25	30	35	40	45
COUNTING										
Forward and Backward	1s, 10s	-5s	+10s		100s		-10s			
Count-On Strategy		from a number								
SYMBOL IDENTIFICATION										
Reading Numbers		teens, 2-digit								
Writing Numbers	1-digit	teens, 2-digit								
PLACE VALUE			10s, 1s addition							
NUMBER FAMILIES			missing number		-0, -all			inverse facts		
Number Family Strategy										
Addition/Subtraction Facts										
Fact Relationships										
Solving Problems Using Number Families										
COLUMN ADDITION AND SUBTRACTION										
Basic Column Addition and Subtraction							no carry/rename			
Column Addition with 3 Addends										
Column Subtraction – Zero as 1st Digit of Answer										
Column Addition with Carrying										
WORD PROBLEMS					write symbols					
Action and Classification Word Problems				more/less						
Comparison Word Problems										
FRACTIONS										
MORE THAN, LESS THAN, EQUAL TO										
Completing Statements										
Associative Property										
TIME										
Digital Clocks										
Analog Clocks										
MONEY										
Coins and Bills									coins	
Writing Dollar Amounts										
Column Problems										
GEOMETRY										
2-Dimensional Shapes										
3-Dimensional Shapes										
Composition/Decomposition										
2-Dimensional Shapes										
3-Dimensional Shapes										

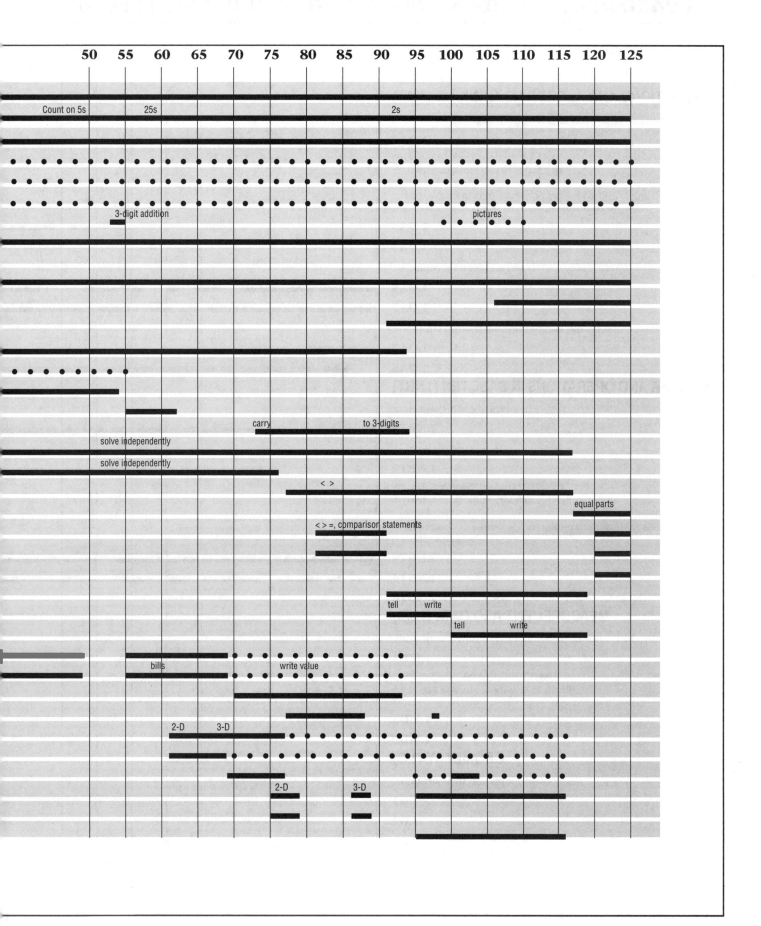

Common Core State Standards Chart and CMC Level B

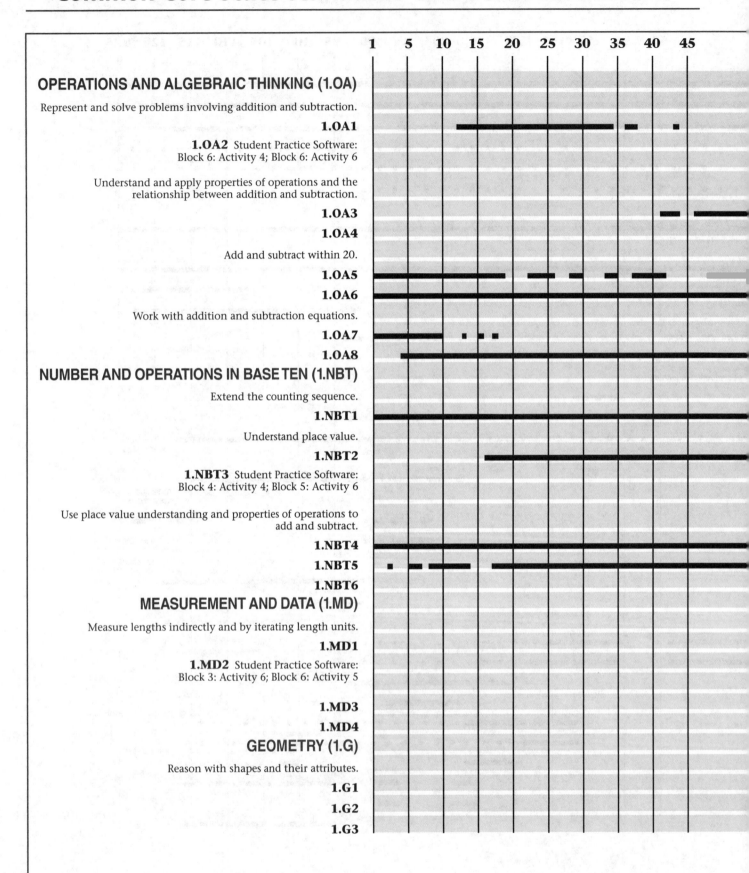

	1	5	10	15	20	25	30	35	40	45

OPERATIONS AND ALGEBRAIC THINKING (1.OA)

Represent and solve problems involving addition and subtraction.

1.OA1

1.OA2 Student Practice Software: Block 6: Activity 4; Block 6: Activity 6

Understand and apply properties of operations and the relationship between addition and subtraction.

1.OA3

1.OA4

Add and subtract within 20.

1.OA5

1.OA6

Work with addition and subtraction equations.

1.OA7

1.OA8

NUMBER AND OPERATIONS IN BASE TEN (1.NBT)

Extend the counting sequence.

1.NBT1

Understand place value.

1.NBT2

1.NBT3 Student Practice Software: Block 4: Activity 4; Block 5: Activity 6

Use place value understanding and properties of operations to add and subtract.

1.NBT4

1.NBT5

1.NBT6

MEASUREMENT AND DATA (1.MD)

Measure lengths indirectly and by iterating length units.

1.MD1

1.MD2 Student Practice Software: Block 3: Activity 6; Block 6: Activity 5

1.MD3

1.MD4

GEOMETRY (1.G)

Reason with shapes and their attributes.

1.G1

1.G2

1.G3

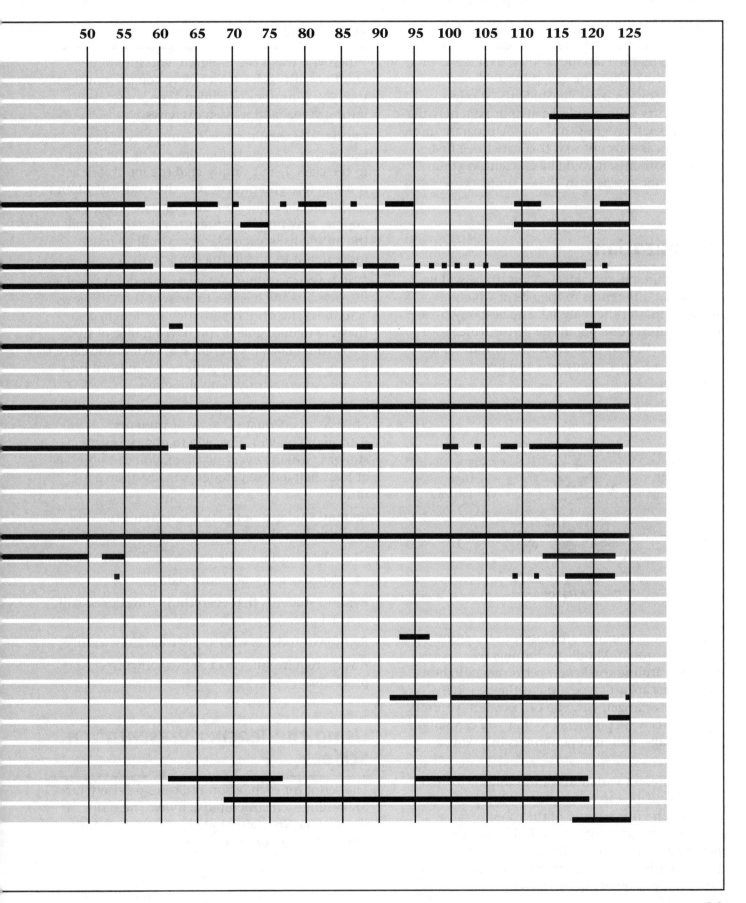

Teaching Effectively

Connecting Math Concepts Level B is designed for students who have the necessary entry skills measured by the Placement Test. (See Placement Tests on page 100.) The group should be as homogeneous as possible. Students who have similar entry skills and learn at approximately the same rate will progress through the program more efficiently as a group. So if there are three first-grade classrooms, it could be efficient to group students homogeneously (based on placement-test scores).

Organization

Even within a homogeneous class, there will be significant differences in the rate at which students master the material. The best way to get timely information about the performance is to arrange seating so you can receive information quickly on higher performers and lower performers.

A good plan is to organize the students something like this:

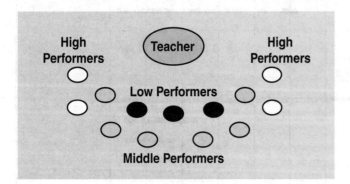

The lowest performers are closest to the front of the classroom. Middle performers are arranged around the lowest performers. Highest performers are arranged around the periphery. With this arrangement, you can position yourself as students work problems so that you can sample low, average, and high performers by taking a few steps.

While different variations of this arrangement are possible, be careful not to seat low performers far from the front center of the room because they require the most feedback. The highest performers, understandably, can be farthest from the center because they attend better, learn faster, and need less observation and feedback.

Teaching

When you teach the program, you should be familiar with each lesson before you present it so that you can monitor students' responses, both during verbal and written exercises.

Ideally, you should rehearse any parts of the lesson that are new before presenting the lesson to the class. Don't simply read the script, but act it out before you present it to the students. Attend to the displays and how the displays change. If you preview the steps students will take to work the problems in each exercise, you'll be much more fluent in presenting the activity.

Watch your wording. Activities that don't involve displays are much easier to present than display activities. The display activities are designed so they are manageable if you have an idea of the steps you'll take. If you rehearse each of the early lessons before presenting them, you'll learn how to present efficiently from the script.

As students work each problem, you should observe an adequate sample of students. Although you won't be able to observe every student working every problem, you can observe at least half a dozen students in less than a minute.

Remind students of the two important rules for doing well in this program:

1. Always work problems the way they are shown.

2. No shortcuts are permitted.

Remind students that everything introduced will be used later.

Reinforce students who apply what they learn.

Always require students to rework incorrect problems.

Using the Teacher Presentation Scripts

The script for each lesson indicates precisely how to present each structured activity. The script shows what you say, what you do, and what the students' responses should be.

What you say appears in blue type:

You say this.

What you do appears in parentheses:

(You do this.)

The responses of the students are in italics.

Students say this.

What you say with students is in blue bold italics.

You and the students say this.

Although you may feel uncomfortable "reading" a script (and you may feel that the students will not pay attention), try to present the exercises as if you're saying something important to the students. If you do, you'll find that working from a script is not difficult and that students respond well to what you say. A sample script appears below and on page 14.

The arrows show six different things you'll do in addition to delivering the wording in the script.

1. You'll **signal, touch or tap,** to make sure group responses involve all the students. (arrows 1a, b, and c)
2. You'll **firm** critical parts of the exercises. (arrows 2a and 2b)
3. You'll **pace** your presentation based on what the students are doing, judging whether to proceed quickly or to wait a few more seconds before moving on with the presentation. (arrows 3a and 3b)
4. You'll **display** (or write) things on the board, and you'll often **add** to the board display. (arrows 4a and 4b)
5. You'll **check** students written work to ensure mastery of the content. (arrow 5)
6. You'll **check** students' mastery with individual turns. (arrow 6)

Example 1

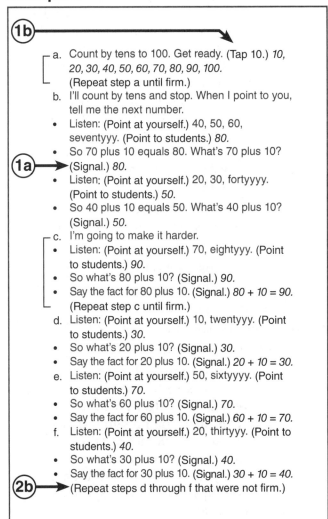

from Lesson 8, Exercise 1

Example 2

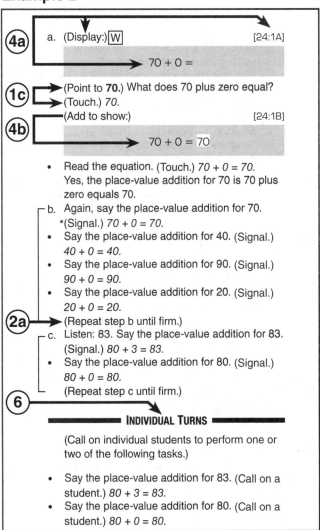

from Lesson 24, Exercise 1

Example 3

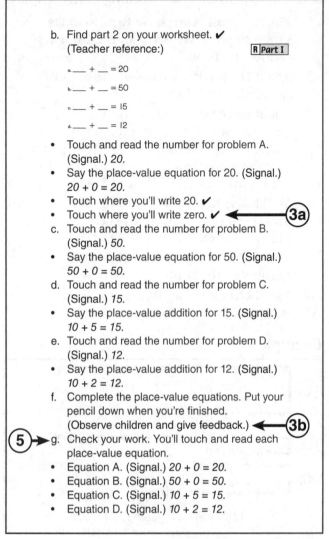

b. Find part 2 on your worksheet. ✔
(Teacher reference:)   R Part I

  a. __ + __ = 20

  b. __ + __ = 50

  c. __ + __ = 15

  d. __ + __ = 12

- Touch and read the number for problem A. (Signal.) *20.*
- Say the place-value equation for 20. (Signal.) *20 + 0 = 20.*
- Touch where you'll write 20. ✔
- Touch where you'll write zero. ✔ ← ③a
c. Touch and read the number for problem B. (Signal.) *50.*
- Say the place-value equation for 50. (Signal.) *50 + 0 = 50.*
d. Touch and read the number for problem C. (Signal.) *15.*
- Say the place-value addition for 15. (Signal.) *10 + 5 = 15.*
e. Touch and read the number for problem D. (Signal.) *12.*
- Say the place-value addition for 12. (Signal.) *10 + 2 = 12.*
f. Complete the place-value equations. Put your pencil down when you're finished. (Observe children and give feedback.) ← ③b
⑤ → g. Check your work. You'll touch and read each place-value equation.
- Equation A. (Signal.) *20 + 0 = 20.*
- Equation B. (Signal.) *50 + 0 = 50.*
- Equation C. (Signal.) *10 + 5 = 15.*
- Equation D. (Signal.) *10 + 2 = 12.*

from Lesson 27, Exercise 4

ARROW 1: SIGNALS

Arrows 1a, 1b, and 1c show the three types of signals teachers use to present tasks in *CMC*. As indicated above, all signals have the same purpose: to trigger a simultaneous response from the group. All signals have the same rationale: If you can get the group to respond simultaneously (with no student leading the others), you will get information about the performance of all the students, not just those who happen to respond first.

The simplest way to signal students to respond together is to adopt a timing practice—just like the timing in a musical piece.

General Rules for Signals

Students will not be able to initiate responses together at the appropriate rate unless you follow these rules:

a. **Talk first.** Pause a standard length of time (possibly one second), then signal. Never signal while you talk. Don't change the timing for your signal. Make your signal predictable, so all students will know when to respond. Students are to respond on your signal—not after it or before it.

b. **Model responses that are paced reasonably.** Don't permit students to produce slow, droning responses. These are dangerous because they rob you of the information that can be derived from appropriate group responses. When students respond in a droning way, many of them are copying responses of others. If students are required to respond at a reasonable speaking rate, all students must initiate responses, and it's relatively easy to determine which students are responding correctly and which students are giving the wrong response. Also, don't permit students to respond at a very fast rate or to "jump" your signal. Listen very carefully to the first part of the response.

To correct mistakes, show the students exactly what you want them to do.

- I'm good at answering the right way.
- Listen. (Point at yourself.) 20, 30, fortyyy. (Touch yourself.) 50.
- Your turn. (Point at yourself.) 20, 30, fortyyy. (Point at students.) 50.

c. **Do not respond with the students unless you are trying to work with them on a difficult response.** You present only what's in blue. You do not say the answers with the students unless the script tells you to. You should not move your lips or give other spurious clues to the answer.

Think of signals this way: If you use them correctly, they provide you with much diagnostic information. A weak response suggests that you should repeat a task and provides information about which students may need more help. Signals are, therefore, important early in the program. After students have learned the routine, the students will be able to respond on cue with no signal. That will happen, however, only if you always give your signals at the end of a constant time interval after you complete what you say.

Basic Signal: Arrow 1a

A basic signal follows a question, a direction, or the words, "Get ready." You can signal for arrow 1a by nodding, dropping your hand, clapping one time, snapping your fingers, or tapping your foot. After initially establishing the timing for signals, you can signal through voice inflection only. Signals specified by the direction (Signal.) maybe be visual or audible, depending on where students' attention is focused. If students are focused on a display or on their Workbook, a basic signal must be audible. If students are focused on you, the basic signal can be audible, but will work if it is only visual (nodding or dropping your hand).

Tap Signal: Arrow 1b

A tap signal is shown by arrow 1b. Signals specified as (Tap.) must be audible signals, such as tapping your foot, snapping your fingers, or clicking a clicker. When students count together, you'll lead them with audible serial taps. The timing of serial taps is extremely important. The timing of the taps for specific tasks should be consistent. If you model a task and then direct students to perform it, the timing of the serial taps should be identical to the rate at which you modeled it.

Point-and-Touch Signal: Arrow 1c

The signal arrow for 1c is a point-and-touch signal. This signal is used in connection with a display. There are two teacher directions for every point-and-touch signal.

Pointing:

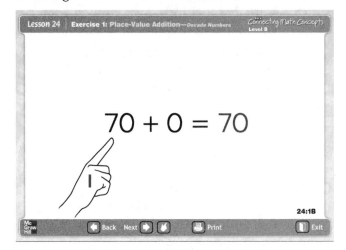

- Hold your finger about an inch from the display, just below what students should focus on.

- Be careful not to cover the material—all the students must be able to see it.

- Hold your finger in the pointing position for at least one second.

- Say the verbal cue for students to respond, "Read the equation."

- Touch elements as students respond to them. (Touch 70, +, 0, =, 70 as students read symbols.) *Seventy plus zero equals seventy*.

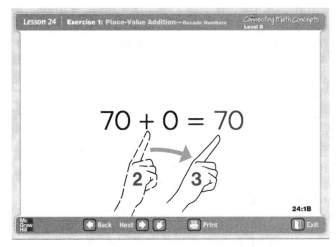

As you touch, try to monitor the lowest performing student in the group to see if the student is responding appropriately.

Common pointing errors:

- Initially touching instead of pointing to what students should focus on
- Pointing to the wrong thing or not clearly pointing to the target material
- Covering the material so students can't see it
- Pointing for less than one second
- Not saying the verbal cue

Common touching errors:

- Touching the symbol before the one-second interval
- Touching the material while you're talking
- Touching the material indecisively or out of rhythm
- Pulling your finger too far from the display making students uncertain what they should be focused on
- Covering the material while touching it
- Failing to monitor low performers

Display CD Signal

The online and CD Board Displays have a special feature that can be used for signaling. The cursor shown on the screen can be replaced by an orange hand icon that, in turn, can be used to signal. You can replace the cursor on the screen by clicking on the icon of a hand at the bottom of the screen. The cursor will turn into an orange hand-shaped cursor. You can move the hand cursor to the part of the display students should focus on. Click to signal, and the hand cursor will move to simulate a "touch" that cues the students' response. You can use the hand cursor to mimic the timing, pointing, and touching of your point-and-touch signal. Practice using the optional hand cursor before using it during a lesson.

- When you click on an exercise, the display codes will be listed for that exercise.
- When you reach the first display in the Teacher Presentation Book, click twice on the appropriate display code to present that display.

ARROW 2: REPEATS

There are two categories of repeats. One kind of repeat alerts you to the importance of the tasks that are taught and implies a special correction if students make a mistake. The other kind of repeat indicates how to present additional examples. Arrow 2b, shows where you may need to make a special correction. Arrow 2a shows the task and the examples that should be repeated.

A special correction is needed when correcting mistakes on tasks that teach a sequence or a relationship. This type of correction is marked with one of three notes.

- (Repeat until firm.) or (Repeat step _____ until firm.) Illustrated by arrow 2a.
- (Repeat steps that were not firm.) Illustrated by arrow 2b.

The repeat directions appear at the end of the tasks that are to be repeated. For repeat-until-firm directions, the tasks and the direction are bracketed on the left to clarify the tasks that need to be repeated. The tasks must be mastered before material that follows is presented.

Repeat-until-firm and repeat-step-_____-until-firm directions are used as follows:

- after a sequence of tasks teaching a relationship that students may not understand if each of the tasks aren't responded to correctly and fluently;
- when students must produce a series of responses in a consistent sequence (as in counting).

For (Repeat until firm.) and (Repeat step _____ until firm.) follow these steps:

1. Correct the mistake. (Tell the answer and repeat the task that was missed.)
2. Return to the beginning of the specified step and present the entire step.

In Example 2, step B, you present four different problems. For each problem, you say, "Say the place-value addition for _____."

If students are confused, if some don't answer a question, or if the answers to any of the questions are wrong, you repeat the bracketed part of step B after you have corrected.

When you hear a mistake, you say the correct answer and repeat the task. However, you make sure that students are firm in all of the problems you present in step B. You cannot be sure that students are firm unless you repeat the step.

Here's a summary of the steps you follow when repeating a part of the exercise when firm.

Correct the mistake:
(Tell the correct answer.)

*   90 plus zero equals 90.

Repeat the task:
*   Say the place-value addition for 90. (Signal.)

Repeat the step:
*   Let's do those again.

(Start at the beginning of the bracket in step B and present the entire task.)

Repeat until firm is based on the information you need about the students. When the students made the mistake, you told the answer. Did they remember the answer? Would they now be able to perform the procedure or sequence of responses correctly? The repeat-until-firm procedure provides you with answers to these questions. Students show you through their responses whether or not the correction worked, whether or not they are firm. Arrow 2a directs you to repeat Example 2, step B until students can say the place-value addition for 70, 40, 90, and 20 without making a mistake. This repeat direction firms a series of responses that must be produced in a consistent sequence.

A **Repeat steps that were not firm** direction occurs when students are expected to apply procedures to a set of examples. Arrow 2b directs you to repeat steps D through F in Example 1, which required you to make a correction. The steps in which errors were made should be repeated until students correctly say the next tens number in the series, answer the plus-10 problem, and say the fact for each step. Only the steps that students miss should be repeated when this direction appears.

ARROW 3: PACING YOUR PRESENTATION AND INTERACTING WITH STUDENTS AS THEY WORK

You should pace your verbal presentation at a normal speaking rate—as if you were telling somebody something important.

Arrows 3a and 3b in Example 3 show two ways to pace your presentation for activities. One is marked with a ✔. The other is a note to **(Observe students and give feedback.).** Both indicate that you will monitor students.

A ✔ is a note to check students' performance on a task that requires only a second or two. If you are positioned close to several lower-performing students, quickly check whether 2 or 3 of them are responding appropriately. If they are, proceed with the presentation.

The **(Observe students and give feedback.)** direction requires more careful observation. You sample more students and you give feedback, not only to individual students but to the group. Here are the basic rules for what to do and what not to do when you observe and give feedback.

a. If the task is one that takes no more than 30 seconds, observe and give feedback to several students. Focus on the lower performers.

b. If the task requires considerably more time, move from the front of the room to a place where you can quickly sample the performance of low, middle, and high performers.

c. As you observe, make comments to the whole class. Focus these comments on students who are following directions, working quickly, and working accurately: "Wow, a couple of students are almost finished. I haven't seen one mistake so far."

d. Students put their pencils down to indicate that they are finished. Acknowledge students who are finished. They are not to work ahead.

e. If you observe mistakes, do *not* provide a great deal of individual help. Point out any mistake, but do not work the problems for the students. For instance, if a student gets one of the problems wrong, point to it and say, "You made a mistake." If students write a wrong number, direct them to "Say the place-value equation for 15." (Signal.) *10 + 5 = 15.* "The equation you wrote for problem C doesn't say 10 plus 5 equals 15. Fix it." Make sure that you check the lower performers and give them feedback. When you show them what they did wrong, keep your explanation simple. The more involved your explanations, the more likely they are to get confused.

f. If you observe a serious problem that is not unique to the lowest performers, tell the class, "Stop. We have a problem." Point out the mistake. Repeat the part of the exercise that gives them information about what they are to do. Do not provide new teaching or new problems. Simply repeat the part of the exercise that gives students the information they need and reassign the work. "Work it the right way."

g. Allow students a reasonable amount of time. Do not wait for the slowest students to complete the problems before presenting the workcheck during which students correct their work and fix any mistakes. You can usually use the middle performers as a gauge for what is reasonable. As you observe that they are completing their work, announce, "Okay, you have about 10 seconds more to finish up." At the end of that time, begin the workcheck.

If you follow the procedures for observing students and giving feedback, your students will work faster and more accurately. They will also become facile at following your directions.

- If you wait a long time period before presenting the workcheck, you punish those who worked quickly and accurately. Soon, they will learn that there is no payoff for doing well—no praise, no recognition—but instead a long wait while you give attention to those who are slow.

- If you don't make announcements about students who are doing well and working quickly, the class will not understand what's expected. Students will probably not improve as much.

- If you provide extensive individual help on written tasks, you will actually reinforce students for not listening to your directions and for being dependent on your help. Furthermore, this dependency becomes contagious. It doesn't take other students in the class long to discover that they don't have to listen to your directions, that they can raise their hand and receive help that shows them how to do the assigned work.

These expectations are the opposite of the ones you want to induce. You want students to be self-reliant and to have reasons for learning and remembering what you say when you instruct them. The simplest reasons are that they will use what they have just been shown and that they will receive reinforcement for performing well.

If you follow the management rules outlined above, all students who are properly placed in the program should be able to complete assigned work within a reasonable period of time and have reasons to feel good about their ability to do math. That's what you want to happen. As students improve, you should tell them about it. "What's this? Everybody's finished with that problem already? That's impressive."

ARROW 4: BOARDWORK/BOARD DISPLAYS CD

In many exercises, you will display problems on the board or a screen.

The word **(Display:)** appears in the script when a display is to be shown to students (arrow 4a). The words **(Add to show:)** appear in the script when something is to be added to an existing display (arrow 4b).

The program has been designed so that you can

a. show all displays and additions or changes to displays by using a CD that comes with the Teacher Presentation Book, or

b. write displays marked with W on the board and project the remaining displays with an overhead projector or a document camera. The displays that are not marked with W are displays that would be difficult to write on a board.

Using the Board Displays CD to Show All Displays

The Board Display CD contains all the displays for every lesson. The displays are labeled consecutively for each lesson. Note that the display code (arrow 4a) shown for each display begins with a number that indicates the lesson. The next number indicates the exercise on that lesson. The letters at the end of the code indicate the order of the displays. The codes for the displays shown in Example 2 are 24:1A and 24:1B. (Example 2 is actually Exercise 1 from Lesson 24.) The 24 indicates Lesson 24. The 1 indicates that it is Exercise 1. The A indicates that it is the first display in that exercise. The B identifies the next display in the exercise. **Note:** The identification code appearing on each CD display corresponds to the code shown in the Presentation Book.

The best way to use the CD is to stand where the images are projected and use a remote device to direct the presentation. (If you are using an interactive white board, you can simply touch the screen.) Being close to the image allows you to point to details of the display as you signal.

Follow these procedures when presenting a new lesson:

- After inserting the disc, double-click on the icon for the Board Displays CD. The computer will display the main menu. Click the desired lesson and then double-click the exercise from the main menu to launch the display. If you select Lesson 24 Exercise 1, the computer screen will show the display for Lesson 24 Exercise 1. From there, you can continue working through all the remaining displays in that lesson without returning to the main menu.

- When you click on an exercise, the display codes will be listed for that exercise.

- When you reach the first display in the Teacher Presentation Book, click twice on the appropriate display code to present that display.

Example 2

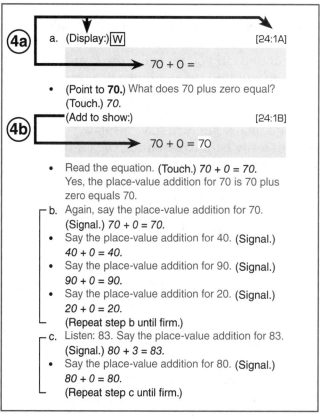

from Lesson 24, Exercise 1

For complicated displays that would take time to write, you can print a copy of the display and use an overhead or document camera to project it.

Here's an example:

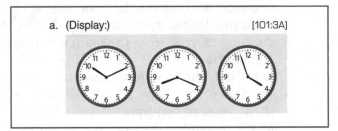

from Lesson 101, Exercise 3

If you aren't going to use a computer for displays, preview lessons to determine which displays you'll write and which will be shown on a document projector.

You can make printed copies of all the displays on the CD by accessing McGraw-Hill/SEG's ConnectED platform. Refer to the card you received with your teacher materials kit for more information.

Whatever system you use, your goal should be to keep the presentation moving without serious interruptions. If you are using the Board Displays CD, you will find that with practice you can present exercises at a good pace. If you are writing some displays on the board and projecting some displays, make sure you have displays you'll project ready before presenting the lesson so you can maintain a good pace.

Advancing CD Displays

You can move from display to display in several ways.

- You can use a remote control. Pressing the forward arrow on the remote calls up the next display. Pressing the back arrow returns to the previous display.

- You can also move from display to display by touching the right or left arrow key on your computer keyboard.

- You can use a mouse or a wireless mouse and click on the on arrows at the bottom of the display.

- For an interactive white board, you can touch the arrows at the bottom of the display.

Other Methods for Presenting Displays

You can teach the program by writing some of the displays on the board. You can copy displays followed by W when they appear in the script.

For example:

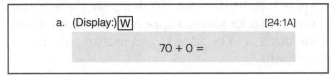

from Lesson 24, Exercise 1

You can write this display on the board. The following display [24:1B] is an add-to-show, which shows changes to a display in white.

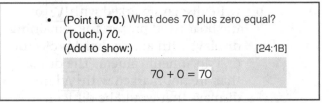

from Lesson 24, Exercise 1

ARROW 5: WORKCHECKS (CHECK YOUR WORK.)

The purpose of the workcheck is to give students timely feedback on their work. It is important for students to correct mistakes on written work, and to do it in a way that allows you to see what mistakes they made.

Students should write their work in pencil so they can erase and make any corrections that are necessary as they work; however, they are not to change their original work once it is finished. The errors that students make provide information on what needs to be firmed or repeated.

During workchecks that involve several problems, circulate among students and check their work. Praise students who are fixing mistakes. Allow a reasonable amount of time for them to check each problem.

Do not wait for the slowest students to finish their check. Try to keep the workcheck moving as quickly as possible.

ARROW 6: INDIVIDUAL TURNS

Individual Turns are specified in the exercises under the heading Individual Turns. Arrow 6 (See page 13.) shows an Individual Turn. Individual Turns usually appear at the end of exercises.

There are several rules to follow when administering individual turns:

1. **Present Individual Turns only after the group is firm.** If you go to Individual Turns too soon, many of the students will not be able to give a firm response. If you wait until the students are firm on group responses, the chances are much better that each will be able to give a firm response on an Individual Turn.

2. **Give most of your Individual Turns to the lowest-performing students.** The lowest performers in the group are those students seated directly in front of you. By watching these students during the group practice of the task, you can tell when they are ready to perform individually. When the lowest performers (who are appropriately placed in the program) can perform the task without further need of correction, you can safely assume that the other students in the group will be able to perform the task. Unless your group is small, do not call on every student, doing so can cause restlessness.

3. **When a student makes a mistake on an Individual Turn, firm the group.** If one individual in the group makes a particular mistake, there is probably one other student in the group who will make the same mistake. The most efficient remedy, therefore, is to firm the entire group. Then retest the student who made the mistake.

If students are consistently weak on Individual Turns, you can assume that students are not mastering the material during your presentation of the tasks to the group. Look for these response problems:

- Group responses not in unison
- Students not attending
- Students repeating mistakes after corrections
- Unmotivated students

The presentation areas that may be causing these problems are:

- Weak signals
- Poor pacing
- Ineffective corrections
- Weak reinforcement

4. **Don't skip Individual Turns.** Always include the Individual Turns for tasks in which they are specified. Individual Turns are not specified in all exercises. If you are in doubt about the performance of any students on a task, present quick individual turns.

INDEPENDENT WORK

The goal of the Independent Work is to provide review of previously taught content. The time required for Independent Work ranges from 10–15 minutes, with the time requirements increasing later in the program. Ideally, all Independent Work is completed in class, but not necessarily during the period in which the structured part of each lesson is presented. If it is not practical for students to complete the work at school, it may be assigned as homework. Students are not to tear out Workbook pages. They need to understand that if they take their Workbook home, they must bring it back the following school day.

Each newly introduced problem type becomes part of the Independent Work after it has appeared several times in structured teacher presentations. Everything that is taught in the program becomes part of the Independent Work. As a general rule, all major problem types that are taught in the program appear at least 10 times in the Independent Work. Some appear as many as 30 or more times. Early material is included in later lessons so that the Independent Work becomes relatively easy for students and provides them with evidence that they are successful.

Unacceptable Error Rates

Students' Independent Work should be monitored, and remedies should be provided for error rates that are too high. As a rule, if more than 30 percent of the students miss more than one or two items in any part of the Independent Work, provide a remedy for that part. The first lesson in which a recently taught skill is independent may have error rates of more than 30 percent of the students. Don't provide a remedy for these situations, but point out to the students that they had trouble with this part and possibly go over the most frequently missed problem. If an excessive error rate continues, provide a systematic correction.

High error rates on independent practice may be the result of the following:

  a. The students may not be placed appropriately in the program.

  b. The initial presentation may not have been adequately firmed. (The students made mistakes that were not corrected. The parts of the teacher presentation in which errors occurred were not repeated until firm.)

  c. Students may have received inappropriate help. (When they worked structured problems earlier, they received too much help and became dependent on the help.)

  d. Students may not have been required to follow directions carefully.

The simplest remedy for unacceptably high error rates on Independent Work is to repeat the structured exercises for that problem type that occurred immediately before the material became independent. For example, if students have an unacceptable error rate on a particular kind of word problem, go to the last one or two exercises that presented the problem type as a teacher-directed activity. Repeat those exercises until students achieve a high level of mastery. Follow the script closely. Make sure you are not providing a great deal of additional prompting. Then assign the Independent Work for which their error rate was too high. Check to make sure students do not make too many errors.

GRADING PAPERS AND FEEDBACK

The teacher material includes a separate Answer Key. The key shows the work for all problems presented during the lesson and as Independent Work. When students are taught a particular method for working problems, they should follow the steps specified in the key. You should make sure students know that the work for a problem is wrong if the procedure is not followed.

After completing each lesson and before presenting the next lesson, follow these steps:

  1. Check for excessive error rates for any parts of the written work from the structured part of the lesson. Note parts that have excessive error rates for more than 30% of the students. For instance, if a particular skill has had high error rates for more than two consecutive days, provide a remedy. Reteach as described above.

  2. Conduct a workcheck for the Independent Work. One procedure is to provide a structured workcheck of Independent Work at the beginning of the period. Do not attempt to provide students with complete information about each problem. Read the answers. Students are to mark each item as correct or incorrect. The workcheck should not take more than five minutes. Students are to correct errors at a later time and hand in the corrected work. Keep records that show for each lesson whether students handed in corrected work. Attend to these aspects of the student's work:

    a. Were all the mistakes corrected?

    b. Is the appropriate work shown for each correction (not just the right answer)?

    c. Did the student perform acceptably on tasks that tended to be missed by other students? The answer to this question provides you with information on the student's performance on difficult tasks.

  3. Award points for Independent Work performance. A good plan is to award one point for completing the Independent Work, one point for correcting all mistakes, and three points for making no more than four errors on the Independent Work. Students who do well can earn five points

for each lesson. These points can be used as part of the basis for assigning grades. The Independent Work should be approximately one-third of the grade. The rest of the grade would be based on the Mastery Tests. (The Independent Work would provide students with up to 50 points for a ten-lesson period; each Mastery Test provides another possible 100 points—100% for a perfect test score.) An online progress monitoring application, *SRA 2Inform*, is available on ConnectED.

INDUCING APPROPRIATE LEARNING BEHAVIORS

Lower Performers

Here are other guidelines for reinforcing appropriate learning behaviors for lower performers who start at Lesson 1.

1. During the first 10 lessons, hold students to a high criterion of performance. Remind them that they are to follow the procedures you show them.

2. If they do poorly on the first Mastery Test, which follows Lesson 10, provide the specified remedies, (See the following section, **In-Program Mastery Tests.**) then repeat Lessons 9 and 10. Tell students, "We're going to do these lessons again. This time, we'll do them perfectly."

Be positive. Reinforce students for following directions and not making the kinds of mistakes they had been making. Understand, however, that for some students, the relearning required to perform well is substantial, so be patient, but persistent. After students have completed lessons with a high level of success, they will understand your criteria for what they should do to perform acceptably. Retest them and point out that their performance shows that they are learning.

Following Directions

Students who have not gone through the earlier level of *CMC* may have strategies for approaching math that are not appropriate. Most notably, they may be very poor at following directions, even if they have an understanding of them. For instance, if you instruct the students to work problem C and then stop, a high percentage of them will not follow this direction.

Throughout *CMC Level B*, you will give students precise directions that may be quite different from those they have encountered earlier in their school experience. The most common problems occur when you direct students to work part of a problem and then stop, work one problem and then stop, or set up a series of problems without solving them.

The simplest remedy is to tell students early in the program that they are to listen to directions and follow them carefully. A good plan is to award points to the group (which means all members of the group earn the same number of points) for following directions. Praise students for attending to them. If you address the issue of following directions early in the school year, students will progress much faster later in the program, and you will not have to nag them as much about following directions.

Exercises that are the most difficult for these students, and the most difficult for the teacher to present effectively, are those that have long explanations. An effective procedure is to move fast enough to keep the students attentive. Go fast on parts that do not require responses from the students; go more slowly and be more emphatic when presenting directions for what the students are to do. If you follow this general guideline, students will attend better to what you say.

Following Solution Steps

An important behavior to address early is that students are to follow the solution steps that are taught. Students work multi-digit addition problems a column at a time beginning on Lesson 27. The procedure students follow is to work the problem in the ones column first, followed by working the problem in the tens column, etc. Students who follow these steps are prepared to solve carrying problems when they are introduced on Lesson 76. Students who do not follow this procedure will have difficulty learning to solve problems that require carrying.

Do not permit shortcuts or working the problem with steps missing. At first, this convention may strike some students as being laborious. Tell them early in the sequence that if they learn these steps, they will later avoid many difficulties students have when trying to work problems that involve a lot of steps. Point out to students that they will learn "real math strategies" that will permit them to have far less difficulty with higher math than they would if they were not well versed in solving problems by following the procedures you teach.

Long Lessons

Expect some lessons to run long. Do not hurry to complete them in one period if it would realistically take more than the period to present and check the material. Complete long lessons during the next period and then start the next lesson during that period and go as far as you can during the allotted time. If you are running long on most of the lessons, the group may not be at mastery, or too much time is being spent on each problem set.

The simplest solution is to select one lesson and repeat it as many as two more times. If the students do not perform rapidly and accurately on all the exercises and work, they are misplaced in the program (which should have been apparent from mastery-test performance). The best solution is to move the group back to an earlier lesson. Administer the previous two Mastery Tests to determine where they should be placed. Then reteach the program starting with the lesson for the new placement. Skip exercises that students have mastered.

In-Program Mastery Tests

Connecting Math Concepts Level B provides 12 in-program tests that permit you to assess how well each student is mastering the program content. The tests are packaged in a separate Student Assessment Book. Tests are scheduled to follow every tenth lesson, starting with Lesson 10 and continuing through Lesson 120. There is no Mastery Test for Lessons 121–125. The primary purpose of the tests is to provide you with information about how well each student is performing on the most recent things that have been taught in the program.

If the information shows that the group did not pass parts of the test, the program provides a specific remedy for each part. The Answer Key shows a passing criterion for each part of the test. A Remedy Table indicates the exercises that are to be repeated before the students are retested. Before presenting the next lesson, provide remedies for parts students fail.

A good method for minimizing the possibility of students copying from each other is to maximize the space between them when they take the test. Discrepancies in the test performance and daily performance of some students pinpoint which students may be copying.

Below is Test 7, which is scheduled after Lesson 70.

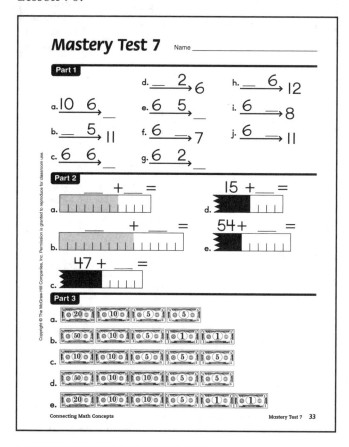

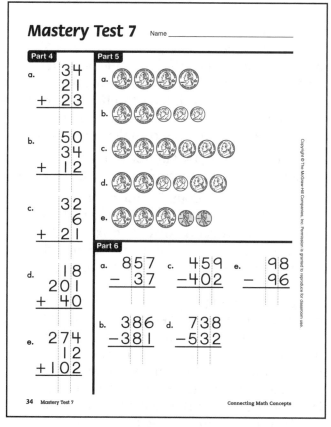

SCORING THE TEST AND PROVIDING REMEDIES

The Answer Key for each test provides the correct answers and shows the work for each item.

Tables that accompany each test show the passing score for each part and indicate the percentages for different total test scores.

Here are the tables for Mastery Test 7:

Passing Criteria Table — Mastery Test 7			
Part	Score	Possible Score	Passing Score
1	2 for each problem	20	18
2	4 for each problem	20	16
3	3 for each problem	15	12
4	3 for each problem	15	12
5	3 for each problem	15	12
6	3 for each problem	15	9
		Total 100	

Passing Criteria Table

The Passing Criteria Table gives the possible points for each item, the possible points for the part, and the passing criterion.

Students fail a part of the test if they score fewer than the specified number of passing points. For example, the total possible points for Part 1 is 20. A passing score is 18. If a student scores 18, 19, or 20, they pass the part. A student with a score of less than 18 fails the part.

Record each student's performance on the Remedy Summary—*Group Summary of Mastery Test Performance* (provided on pages 119–122 of this Teacher's Guide). The Group Summary accommodates up to 30 students. The sample on the next page shows the results for only 6 students.

Here's the results for 6 children on Mastery Test 7:

Remedy Summary—Group Summary of Test Performance

Note: Test remedies are specified in the Answer Key.

Name	1	2	3	4	5	6	Total %
1. Amanda			✔	✔		✔	79%
2. Karen				✔			91%
3. Adam		✔		✔	✔	✔	71%
4. Chan							89%
5. Felipe		✔					95%
6. Jack				✔		✔	83%
Number of students Not Passed = NP	0	2	1	4	1	3	
Total number of students = T	6	6	6	6	6	6	
Remedy needed if NP/T = 25% or more	N	Y	N	Y	N	Y	

(Columns 1–6 are headed "Check parts not passed," under the title "Test 7")

Remedy Summary, Group Summary

The summary sheet provides you with a cumulative record of each student's performance on the in-program Mastery Tests.

Summarize each student's performance by making a check mark for each part failed.

At the bottom of each column, write the total number of failures for that part and the total number of students in the class. Then divide the number of failures by the number of students to determine the failure rate.

Provide a group remedy for each part that has a failure rate of more than 25% (.25).

Test Remedies

The Answer Key specifies remedies for each test. Any necessary remedies should be presented before the next lesson (Lesson 71.)

Here are the remedies for Test 7:

Remedy Table — Mastery Test 7

Part	Test Items	Remedy Lesson	Remedy Exercise	Student Material Remedies Worksheet
1	New Number Families	64	1	—
		65	2	—
		67	1	—
2	Count On (Rulers)	60	3	—
		63	7	Part A
		68	8	Part B
		69	7	Part C
3	Bills	59	2	—
		60	4	—
		62	8	Part D
		64	8	Part E
4	3 Addends (Columns)	64	7	Part F
		65	7	Part G
5	Coins (Quarters)	61	6	Part H
		62	6	Part I
		64	5	Part J
6	Column Subtraction	62	2	—

Remedy Table

If the same students frequently fail parts of the test, it may be possible to provide remedies for those students as the others do Independent Work. If individual students are weak on a particular skill, they will have trouble later in the program when that skill becomes a component in a larger operation or more complex application.

If students consistently fail tests, they are probably not placed appropriately in the program.

On the completed Group Summary of Test Performance for Mastery Test 7, more than one-quarter of the students failed Parts 2, 4, and 6. You provide group remedies by re-teaching the exercises specified as the remedy for those parts. The table indicates to present 4 exercises as remedies for failing Part 2 (Count on). The first remedy exercise is from Lesson 60, Exercise 3. The hyphen in the column for Student Material indicates that there is no Student Material needed to present this exercise. The second remedy for Part 2 of the test is Lesson 63, Exercise 7. A reproducible copy of the Student Material for this exercise can be found on the Remedy Worksheet. (See next page.)

The summary indicates that you may also need to provide individual remedies for Amanda and Adam because they failed additional parts.

It may not be possible or practical to follow all the steps indicated for correcting individual students who make too many mistakes. For these cases, provide test remedies to the *entire class* and move on in the program, attending to those students who make chronic mistakes without significantly slowing the group's progress.

Remedy Worksheets

All the Workbook parts needed for remedies appear in the Student Assessment Book immediately after each test. These are the Remedy worksheets. Below are Remedy parts A through D for Test 7, Parts 2 through 5. These parts are identical to the Workbook parts for the exercises. They are reproduced in the Student Assessment Book so students have a clean copy to work on.

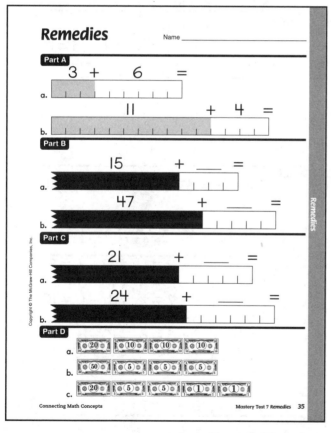

from Remedy Worksheet

Cumulative Tests

CMC Level B has two Cumulative Tests. The first appears after Mastery Test 6 in the Student Assessment Book. The other Cumulative Test appears after Lesson 125, at the end of the program.

Note that you don't have to present the Cumulative Test immediately after students complete Mastery Test 6. You may present Cumulative Test 1 any time after students reach Lesson 60 in the program. You may present Cumulative Test 2 after students reach the end of the program (Lesson 125).

These tests sample critical skills and concepts that have been taught earlier. The first Cumulative Test assesses content from the beginning of the program through Lesson 60.

The final Cumulative Test samples critical items from the beginning of the level, but it is heavily weighted for the lesson range from 61 to the end of the program.

The presentation scripts for the Cumulative Tests appear near the middle of Teacher Presentation Book 2 and near the end of Teacher Presentation Book 3. Cumulative Test 1 is on pages 131–136 in Presentation Book 2 (mid-program test) and Cumulative Test 2 is on pages 342–353 in Presentation Book 3 (final test). These tests usually require more than one period to administer.

Test remedies for Cumulative Test 1 and Cumulative Test 2 are available on ConnectED as well as remedy worksheets.

The Answer Keys for Cumulative Tests 1 and 2 follow.

Cumulative Test 1 Name _____

Part 1
a. 18 b. 14 c. 15 d. 90 e. 80
f. 20 g. 42 h. 53 i. 27 j. 13

Part 2

	i. 4+1=5	r. 7+2=9
a. 2+1=3	j. 14+1=15	s. 10+1=11
b. 12+1=13	k. 8+1=9	t. 10+2=12
c. 6+1=7	l. 8+2=10	u. 4+0=4
d. 16+1=17	m. 9+1=10	v. 1+0=1
e. 3+1=4	n. 9+2=11	w. 10+0=10
f. 13+1=14	o. 5+1=6	x. 38+0=38
g. 1+1=2	p. 5+2=7	y. 0+0=0
h. 11+1=12	q. 7+1=8	

Part 3

90 + 6 = 96

a. 30 + 7 = 37   d. 10 + 9 = 19
b. 80 + 1 = 81   e. 10 + 5 = 15
c. 90 + 0 = 90   f. 50 + 0 = 50

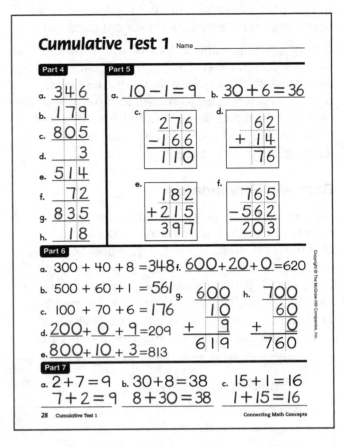

Cumulative Test 1 Name _____

Part 4
a. 346
b. 179
c. 805
d. 3
e. 514
f. 72
g. 835
h. 18

Part 5
a. 10 − 1 = 9   b. 30 + 6 = 36
c. 276 − 166 = 110   d. 62 + 14 = 76
e. 182 + 215 = 397   f. 765 − 562 = 203

Part 6
a. 300 + 40 + 8 = 348   f. 600 + 20 + 0 = 620
b. 500 + 60 + 1 = 561   g. 600 + 10 + 9 = 619
c. 100 + 70 + 6 = 176   h. 700 + 60 + 0 = 760
d. 200 + 0 + 9 = 209
e. 800 + 10 + 3 = 813

Part 7
a. 2+7=9   b. 30+8=38   c. 15+1=16
   7+2=9   8+30=38   1+15=16

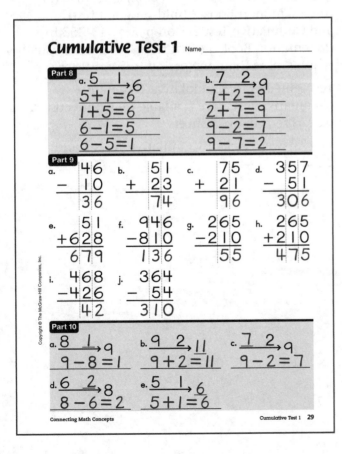

Cumulative Test 1 Name _____

Part 8
a. 5 1 → 6
5+1=6
1+5=6
6−1=5
6−5=1

b. 7 2 → 9
7+2=9
2+7=9
9−2=7
9−7=2

Part 9
a. 46 − 10 = 36
b. 51 + 23 = 74
c. 75 + 21 = 96
d. 357 − 51 = 306
e. 51 + 628 = 679
f. 946 − 810 = 136
g. 265 − 210 = 55
h. 265 + 210 = 475
i. 468 − 426 = 42
j. 364 − 54 = 310

Part 10
a. 8 1 → 9   9−8=1
b. 9 2 → 11   9+2=11
c. 7 2 → 9   9−2=7
d. 6 2 → 8   8−6=2
e. 5 1 → 6   5+1=6

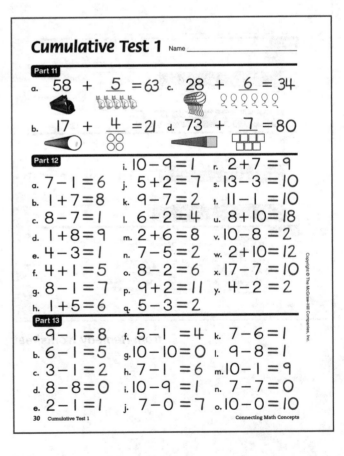

Cumulative Test 1 Name _____

Part 11
a. 58 + 5 = 63   c. 28 + 6 = 34
b. 17 + 4 = 21   d. 73 + 7 = 80

Part 12

	i. 10−9=1	r. 2+7=9
a. 7−1=6	j. 5+2=7	s. 13−3=10
b. 1+7=8	k. 9−7=2	t. 11−1=10
c. 8−7=1	l. 6−2=4	u. 8+10=18
d. 1+8=9	m. 2+6=8	v. 10−8=2
e. 4−3=1	n. 7−5=2	w. 2+10=12
f. 4+1=5	o. 8−2=6	x. 17−7=10
g. 8−1=7	p. 9+2=11	y. 4−2=2
h. 1+5=6	q. 5−3=2	

Part 13
a. 9−1=8   f. 5−1=4   k. 7−6=1
b. 6−1=5   g. 10−10=0   l. 9−8=1
c. 3−1=2   h. 7−1=6   m. 10−1=9
d. 8−8=0   i. 10−9=1   n. 7−7=0
e. 2−1=1   j. 7−0=7   o. 10−0=10

Cumulative Test 1 Name _____

Part 14
a. 🪙🪙🪙🪙🪙 = 50
b. 🪙🪙🪙🪙🪙 = 25
c. 🪙🪙🪙🪙🪙🪙 = 42
d. 🪙🪙🪙🪙🪙🪙🪙 = 48

Part 15
a. $70+10=80$ e. $50+10=60$
b. $72+10=82$ f. $53+10=63$
c. $20+10=30$ g. $30+10=40$
d. $25+10=35$ h. $36+10=46$

Part 16
a. $1+7+2=10$ c. $1+4+2=7$
b. $9+1+6=16$ d. $1+9+5=15$

Part 17
d. $9-3=6$ h. $3+6=9$
a. $8-3=5$ e. $6+4=10$ i. $10-4=6$
b. $10-6=4$ f. $9-6=3$ j. $5+3=8$
c. $3+5=8$ g. $8-5=3$

Connecting Math Concepts   Cumulative Test 1   31

Cumulative Test 2

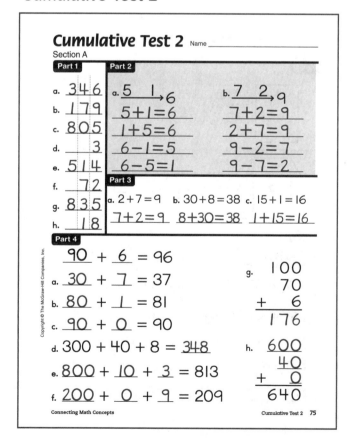

Cumulative Test 2 Name _____
Section A

Part 1
a. 346
b. 179
c. 805
d. 3
e. 514
f. 72
g. 835
h. 18

Part 2
a. $\dfrac{5\ 1}{} \to 6$
$5+1=6$
$1+5=6$
$6-1=5$
$6-5=1$

b. $\dfrac{7\ 2}{} \to 9$
$7+2=9$
$2+7=9$
$9-2=7$
$9-7=2$

Part 3
a. $2+7=9$ b. $30+8=38$ c. $15+1=16$
$7+2=9$ $8+30=38$ $1+15=16$

Part 4
$90 + 6 = 96$
a. $30 + 7 = 37$
b. $80 + 1 = 81$
c. $90 + 0 = 90$
d. $300 + 40 + 8 = 348$
e. $800 + 10 + 3 = 813$
f. $200 + 0 + 9 = 209$

g.
$$\begin{array}{r} 100 \\ 70 \\ +\ \ 6 \\ \hline 176 \end{array}$$

h.
$$\begin{array}{r} 600 \\ 40 \\ +\ \ 0 \\ \hline 640 \end{array}$$

Connecting Math Concepts   Cumulative Test 2   75

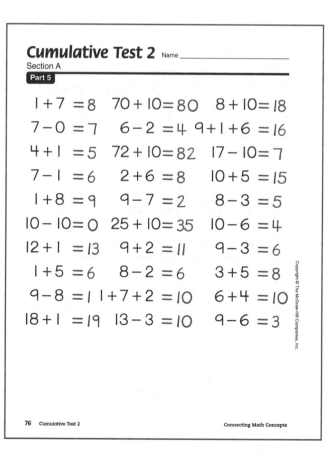

Cumulative Test 2 Name _____
Section A

Part 5

$1+7=8$	$70+10=80$	$8+10=18$
$7-0=7$	$6-2=4$	$9+1+6=16$
$4+1=5$	$72+10=82$	$17-10=7$
$7-1=6$	$2+6=8$	$10+5=15$
$1+8=9$	$9-7=2$	$8-3=5$
$10-10=0$	$25+10=35$	$10-6=4$
$12+1=13$	$9+2=11$	$9-3=6$
$1+5=6$	$8-2=6$	$3+5=8$
$9-8=1$	$1+7+2=10$	$6+4=10$
$18+1=19$	$13-3=10$	$9-6=3$

76   Cumulative Test 2   Connecting Math Concepts

Connecting Math Concepts   *Teacher's Guide*   **29**

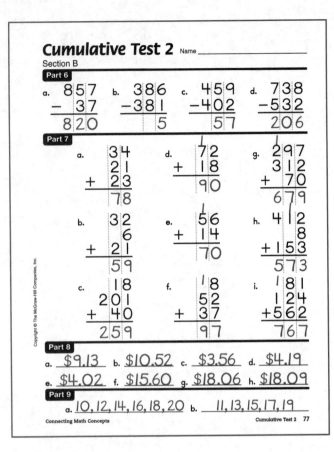

Cumulative Test 2 Name _____

Section B

Part 6

a. 857 − 37 = 820
b. 386 − 381 = 5
c. 459 − 402 = 57
d. 738 − 532 = 206

Part 7

a. 34 + 21 + 23 = 78
b. 32 + 6 + 21 = 59
c. 18 + 201 + 40 = 259
d. 72 + 18 = 90
e. 56 + 14 + 70 = 70
f. 8 + 52 + 37 = 97
g. 297 + 312 + 70 = 679
h. 412 + 8 + 153 = 573
i. 181 + 124 + 562 = 767

Part 8

a. $9.13 b. $10.52 c. $3.56 d. $4.19
e. $4.02 f. $15.60 g. $18.06 h. $18.09

Part 9

a. 10, 12, 14, 16, 18, 20   b. 11, 13, 15, 17, 19

Connecting Math Concepts   Cumulative Test 2 77

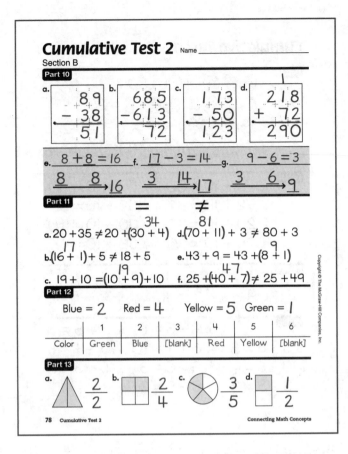

Cumulative Test 2 Name _____

Section B

Part 10

a. 89 − 38 = 51
b. 685 − 613 = 72
c. 173 − 50 = 123
d. 218 + 72 = 290

e. 8 + 8 = 16 → 8 8 → 16
f. 17 − 3 = 14 → 3 14 → 17
g. 9 − 6 = 3 → 3 6 → 9

Part 11   =   ≠

a. 20 + 35 ≠ 20 + (30 + 4)   [34]
b. (16 + 1) + 5 ≠ 18 + 5   [17]
c. 19 + 10 = (10 + 9) + 10   [19]
d. (70 + 11) + 3 ≠ 80 + 3   [81]
e. 43 + 9 = 43 + (8 + 1)   [9]
f. 25 + (40 + 7) ≠ 25 + 49   [47]

Part 12

Blue = 2   Red = 4   Yellow = 5   Green = 1

	1	2	3	4	5	6
Color	Green	Blue	[blank]	Red	Yellow	[blank]

Part 13

a. 2/2   b. 2/4   c. 3/5   d. 1/2

78 Cumulative Test 2   Connecting Math Concepts

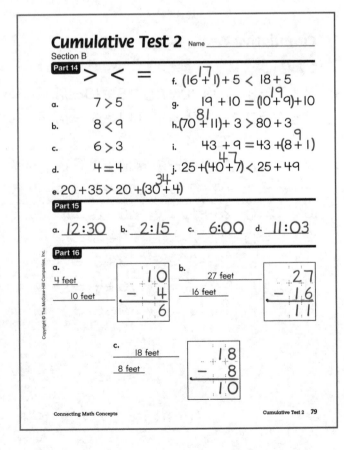

Cumulative Test 2 Name _____

Section B

Part 14   > < =

a. 7 > 5
b. 8 < 9
c. 6 > 3
d. 4 = 4
e. 20 + 35 > 20 + (30 + 4)   [34]
f. (16 + 1) + 5 < 18 + 5   [17]
g. 19 + 10 = (10 + 9) + 10   [19]
h. (70 + 11) + 3 > 80 + 3   [81]
i. 43 + 9 = 43 + (8 + 1)   [9]
j. 25 + (40 + 7) < 25 + 49   [47]

Part 15

a. 12:30   b. 2:15   c. 6:00   d. 11:03

Part 16

a. 4 feet / 10 feet   10 − 4 = 6
b. 27 feet / 16 feet   27 − 16 = 11
c. 18 feet / 8 feet   18 − 8 = 10

Connecting Math Concepts   Cumulative Test 2 79

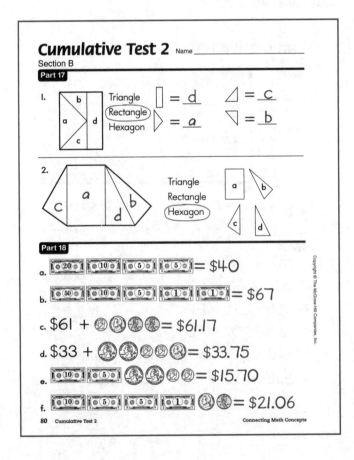

Cumulative Test 2 Name _____

Section B

Part 17

1. Triangle = d   Rectangle = a   Hexagon = b
   △ = c   ▽ = b

2. Hexagon

Triangle
Rectangle
Hexagon

Part 18

a. = $40
b. = $67
c. $61 + = $61.17
d. $33 + = $33.75
e. = $15.70
f. = $21.06

80 Cumulative Test 2   Connecting Math Concepts

Connecting Math Concepts

Cumulative Test 2 Name_____

Section B

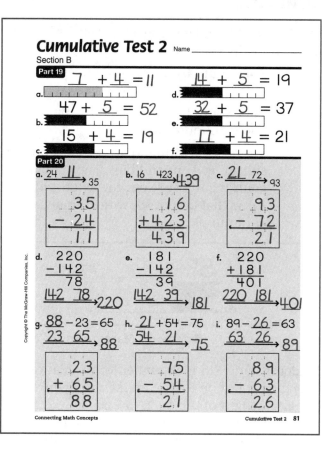

Part 19

a. $\underline{7} + \underline{4} = 11$
d. $\underline{14} + \underline{5} = 19$

b. $47 + \underline{5} = 52$
e. $\underline{32} + \underline{5} = 37$

c. $15 + \underline{4} = 19$
f. $\underline{17} + 4 = 21$

Part 20

a. $24 \quad \underline{11} \rightarrow 35$

$$\begin{array}{r} 3.5 \\ -2.4 \\ \hline 1\,1 \end{array}$$

b. $16 \quad 423 \rightarrow \underline{439}$

$$\begin{array}{r} 1.6 \\ +4.23 \\ \hline 439 \end{array}$$

c. $21 \quad 72 \rightarrow 93$

$$\begin{array}{r} 9.3 \\ -7.2 \\ \hline 2\,1 \end{array}$$

d. $\begin{array}{r} 220 \\ -142 \\ \hline 78 \end{array}$

$\underline{142} \quad 78 \rightarrow 220$

e. $\begin{array}{r} 181 \\ -142 \\ \hline 39 \end{array}$

$\underline{142} \quad 39 \rightarrow 181$

f. $\begin{array}{r} 220 \\ +181 \\ \hline 401 \end{array}$

$220 \quad 181 \rightarrow \underline{401}$

g. $\underline{88} - 23 = 65$

$23 \quad 65 \rightarrow 88$

$$\begin{array}{r} 2.3 \\ +6.5 \\ \hline 88 \end{array}$$

h. $\underline{21} + 54 = 75$

$54 \quad 21 \rightarrow \underline{75}$

$$\begin{array}{r} 7.5 \\ -5.4 \\ \hline 2\,1 \end{array}$$

i. $89 - \underline{26} = 63$

$63 \quad 26 \rightarrow 89$

$$\begin{array}{r} 8.9 \\ -6.3 \\ \hline 2\,6 \end{array}$$

Connecting Math Concepts   Cumulative Test 2 **81**

Cumulative Test 2 Name_____

Section B

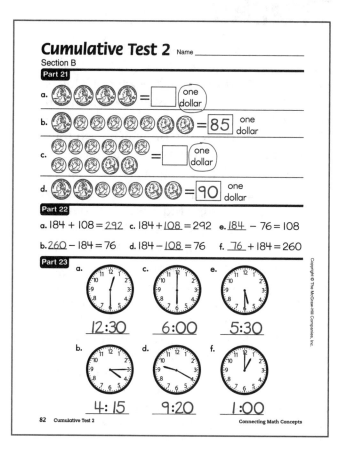

Part 21

a. [coins] $=$ [ ] one dollar

b. [coins] $= \boxed{85}$ one dollar

c. [coins] $=$ [ ] one dollar

d. [coins] $= \boxed{90}$ one dollar

Part 22

a. $184 + 108 = \underline{292}$ c. $184 + \underline{108} = 292$ e. $\underline{184} - 76 = 108$

b. $\underline{260} - 184 = 76$ d. $184 - \underline{108} = 76$ f. $\underline{76} + 184 = 260$

Part 23

a. [clock] $12:30$ c. [clock] $6:00$ e. [clock] $5:30$

b. [clock] $4:15$ d. [clock] $9:20$ f. [clock] $1:00$

82 Cumulative Test 2   Connecting Math Concepts

Cumulative Test 2 Name_____

Section B

Part 24

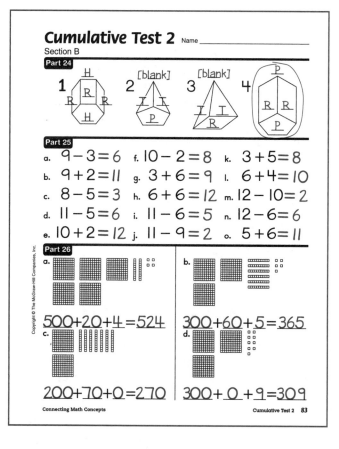

1 [hexagon: H, R, R, H] 2 [blank] 3 [blank] 4 [P, R, R, P]

Part 25

a. $9 - 3 = 6$ f. $10 - 2 = 8$ k. $3 + 5 = 8$

b. $9 + 2 = 11$ g. $3 + 6 = 9$ l. $6 + 4 = 10$

c. $8 - 5 = 3$ h. $6 + 6 = 12$ m. $12 - 10 = 2$

d. $11 - 5 = 6$ i. $11 - 6 = 5$ n. $12 - 6 = 6$

e. $10 + 2 = 12$ j. $11 - 9 = 2$ o. $5 + 6 = 11$

Part 26

a. [base-ten blocks] $500 + 20 + 4 = 524$

b. [base-ten blocks] $300 + 60 + 5 = 365$

c. [base-ten blocks] $200 + 70 + 0 = 270$

d. [base-ten blocks] $300 + 0 + 9 = 309$

Connecting Math Concepts   Cumulative Test 2 **83**

Cumulative Test 2 Name_____

Section B

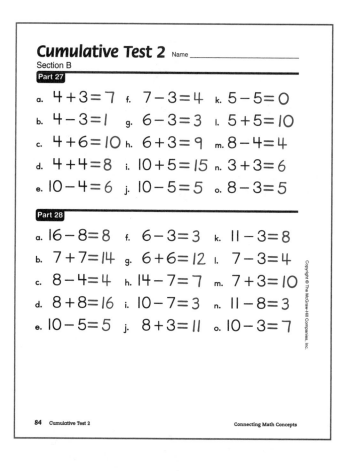

Part 27

a. $4 + 3 = 7$ f. $7 - 3 = 4$ k. $5 - 5 = 0$

b. $4 - 3 = 1$ g. $6 - 3 = 3$ l. $5 + 5 = 10$

c. $4 + 6 = 10$ h. $6 + 3 = 9$ m. $8 - 4 = 4$

d. $4 + 4 = 8$ i. $10 + 5 = 15$ n. $3 + 3 = 6$

e. $10 - 4 = 6$ j. $10 - 5 = 5$ o. $8 - 3 = 5$

Part 28

a. $16 - 8 = 8$ f. $6 - 3 = 3$ k. $11 - 3 = 8$

b. $7 + 7 = 14$ g. $6 + 6 = 12$ l. $7 - 3 = 4$

c. $8 - 4 = 4$ h. $14 - 7 = 7$ m. $7 + 3 = 10$

d. $8 + 8 = 16$ i. $10 - 7 = 3$ n. $11 - 8 = 3$

e. $10 - 5 = 5$ j. $8 + 3 = 11$ o. $10 - 3 = 7$

84 Cumulative Test 2   Connecting Math Concepts

Tracks

This section provides more detail about the skills and knowledge students learn in *CMC Level B*. The skills are divided into tracks. Each track focuses on specific skills. The **Scope and Sequence Chart**, on pages 8–9, shows the tracks of *CMC Level B*. The bars show the lesson on which each track begins and the lesson range for the track.

Each lesson is composed of exercises from 6 to 9 tracks that present various (sometimes unrelated) skills. In other words, students do not devote a lesson to a single topic or skill. Rather they spend a few minutes working on each exercise.

As component skills are learned, they merge into operations. For example, students learn to identify columns as hundreds, tens, and ones, and learn to write digits in the appropriate columns. Separately, they learn to add two numbers and later to add three numbers, for example, 1 + 4 + 2. Finally, students learn how to carry. The procedure uses what they have learned about columns and adding. They write 2-digit sums so the ones digit is in the ones column and the tens digit in the tens column.

For example:

$$\begin{array}{r} {\scriptstyle 1} \\ 4\,7 \\ +\,2\,3 \\ \hline 0 \end{array}$$

See **CMC Level B and Common Core State Standards for Mathematics** for a discussion about how *CMC Level B* meets some of the standards specified for Grade 1 by the Common Core State Standards for Mathematics that are not addressed in this section.

Counting

In *CMC Level B* students learn a variety of counting skills.

- They count by different numbers—ones, tens, hundreds, fives, 25s, twos.

- They count backward by ones, by tens, and by hundreds.

- They use counting to determine amounts shown by coins and bills. They also use counting to figure out the lengths of rulers that do not show the initial units. For example:

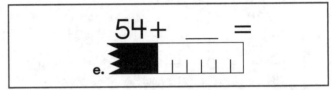

Workbook, Mastery Test 7, Part 2 E

Students start with the number for the black part (54) then count on for the markers to figure out the total length of the ruler. (**54,** 55, 56, 57, 58, 59, 60.) Similar problems show the end number, but no numbers for the black part. Students count backward to figure out the length of the black part.

One important role of counting in the early lessons is to establish the basis for adding and subtracting. In Lesson 1, counting by ones is first reviewed, then it is established as the basis for adding 1. Counting backward is introduced on Lesson 4. Students are then shown the relationship between subtracting 1 and counting backwards.

COUNTING FORWARD AND BACKWARD

Counting by Ones

Here's the first counting exercise from Lesson 1. Students count by ones to ten, to twenty, and to numbers between 10 and 20.

EXERCISE 2: COUNT BY ONES

a. My turn to count to 10: 1, 2, 3, 4, 5, 6, 7, 8, 9, 10.
- Your turn: Count to 10. Get ready. (Tap 10.) *1, 2, 3, 4, 5, 6, 7, 8, 9, 10.*
  (Repeat step a until firm.)

b. My turn to start with 3 and count to 10: 3, 4, 5, 6, 7, 8, 9, 10.
- Your turn: Start with 3 and count to 10. Get ready. (Tap 8.) *3, 4, 5, 6, 7, 8, 9, 10.*

c. Start with 5 and count to 10. Get ready. (Tap 6.) *5, 6, 7, 8, 9, 10.*
- Start with 2 and count to 10. Get ready. (Tap 9.) *2, 3, 4, 5, 6, 7, 8, 9, 10.*
  (Repeat steps b and c until firm.)

d. My turn to start with 10 and count to 20.
- What will I start with? (Signal.) *10.*
- What will I count to? (Signal.) *20.*
- Here I go: 10, 11, 12, 13, 14, 15, 16, 17, 18, 19, 20.

e. Your turn: Start with 10 and count to 20. Get ready. (Tap 11.) *10, 11, 12, 13, 14, 15, 16, 17, 18, 19, 20.*

f. My turn to start with 10 and count to 15: 10, 11, 12, 13, 14, 15.
- Your turn: Start with 10 and count to 15. Get ready. (Tap 6.) *10, 11, 12, 13, 14, 15.*
- Start with 11 and count to 15. Get ready. (Tap 5.) *11, 12, 13, 14, 15.*

g. My turn to start with 15 and count to 20: 15, 16, 17, 18, 19, 20.
- Your turn: Start with 15 and count to 20. Get ready. (Tap 6.) *15, 16, 17, 18, 19, 20.*
- Your turn: Start with 17 and count to 20. Get ready. (Tap 4.) *17, 18, 19, 20.*
- Your turn: Start with 14 and count to 20. Get ready. (Tap 7.) *14, 15, 16, 17, 18, 19, 20.*
  (Repeat steps d through g until firm.)

Lesson 1, Exercise 2

Teaching Note: All the students are able to count to 20. The purpose of this exercise is to review the counting and to provide students with practice in producing proper unison responses.

You model some tasks before the students do the task. Make sure that you model at the same rate that you expect students to respond. Stop them as soon as it is apparent that not all students are responding together and at the right rate.

The correction is to model again, direct the students to count with you, and then direct the students to count without your lead. A good practice is to tap in time with your counting; then tap at the same rate when the students count.

If necessary, repeat the counting three or more times, but don't make it seem like punishment. Make comments like "Now you're getting it," and act as if it is fun to respond together.

Counting by Tens

In Lesson 1, students also count by tens. The teacher shows that the tens numbers for 10 through 100 are created by adding zero to the ones numbers 1–10.

Here's the exercise:

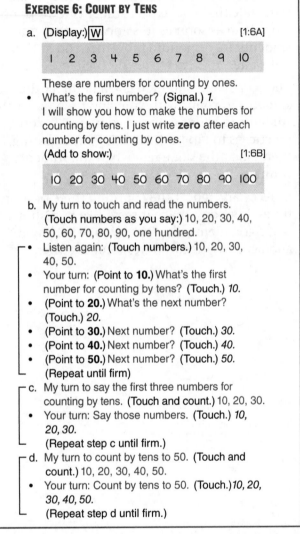

Lesson 1, Exercise 6

On the following lessons, students receive further practice in counting by tens.

Counting Backward by Ones

Counting backward is introduced on Lesson 4. The numbers are shown on a number line. Students start with a larger number and then move backward along the line as they read the numbers.

Here's part of the exercise from Lesson 4:

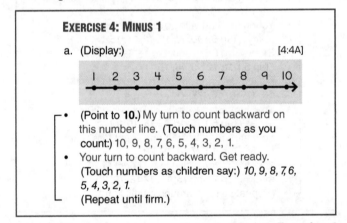

from Lesson 4, Exercise 4

Counting backwards with 2-digit numbers is introduced on Lesson 16. The students first count backward from 7 to 1; then the teacher models counting backward from 37 to 31. Students count from 37 to 31, then from 87 and 27.

Here's part of the exercise from Lesson 16:

EXERCISE 7: MIXED COUNTING
COUNTING BACKWARD

a. Now we'll count backward. I'll start with 7 and count backward. Sevennn, 6, 5, 4, 3, 2, 1.
   • Your turn: Start with 7 and count backward. Get 7 going. *Sevennnn*. Count backward. (Tap 6.) *6, 5, 4, 3, 2, 1.*
   (Repeat until firm.)

b. I'll start with 37 and count backward to 31.
   • What am I going to start with? (Signal.) *37.* Thirty-sevennn, 36, 35, 34, 33, 32, 31.
   • Your turn: Start with 37 and count backward to 31. Get 37 going. *Thirty-sevennn*. Count backward. (Tap 6.) *36, 35, 34, 33, 32, 31.*
   (Repeat step b until firm.)

c. Your turn: Start with 87 and count backward to 81.
   • Get 87 going. *Eighty-sevennn*. Count backward. (Tap 6.) *86, 85, 84, 83, 82, 81.*
   (Repeat step c until firm.)

d. Your turn: Start with 27 and count backward to 21.
   • Get 27 going. *Twenty-sevennn*. Count backward. (Tap 6.) *26, 25, 24, 23, 22, 21.*
   (Repeat step d until firm.)

from Lesson 16, Exercise 7

Teaching Note: For these series, stress the ones digits when you model the counting:

Thirty SEVEN, thirty SIX . . .

Later in this track, students count series that go from one decade to the one that precedes it: for example counting backward from 42 to 38.

This work begins on Lesson 23. Students first count backward from 40 to 30.

Here is part of the exercise from Lesson 23:

EXERCISE 6: MIXED COUNTING

a. You're going to start with 40 and count backward to 30. My turn to do the hard part. Fortyyy, 39, 38.
   • Your turn to say the hard part. Get 40 going. *Fortyyy.* Count backward. (Tap 2.) *39, 38.*

b. You're going to start with 40 and count backward to 30. Get 40 going. *Fortyyy.* Count backward. (Tap 10.) *39, 38, 37, 36, 35, 34, 33, 32, 31, 30.* (Repeat step b until firm.)

c. This time I won't tell you the hard part. You're going to start with 80 and count backward to 70. Get 80 going. *Eightyyy.* Count backward. (Tap 10.) *79, 78, 77, 76, 75, 74, 73, 72, 71, 70.*

d. You're going to start with 30 and count backward to 20. Get 30 going. *Thirtyyy.* Count backward. (Tap 10.) *29, 28, 27, 26, 25, 24, 23, 22, 21, 20.*

e. You're going to start with 70 and count backward to 60. Get 70 going. *Seventyyy.* Count backward. (Tap 10.) *69, 68, 67, 66, 65, 64, 63, 62, 61, 60.*
   (Repeat steps c through e until firm.)

from Lesson 23, Exercise 6

Teaching Note: The most difficult number for students is the number after the tens number (40, 80, 30, 60).

You model the hard part in step A. If students have trouble with the following sequences, tell them you're counting from 80 to 70. So all the numbers you say when you count will be 70s numbers.

   • What's the biggest 70s number?
   • That's the number you say first when you count backward from 80.

COUNTING-ON

On Lesson 5, students are introduced to the count-on strategy that they use throughout the program.

For this strategy they get the starting number going by holding it for at least a second. If the starting number is 4, they say *fouuur* to get it going.

Then they count from 4 at the regular counting pace, *5, 6, 7, 8.*

Here is the introduction to counting-on:

EXERCISE 6: COUNT BY ONES
GET IT GOING

a. I'll show you how to get numbers going. Here's how I get 4 going: Fouuur.
• Your turn: Get 4 going. (Signal.) *Fouuur.*
b. My turn to get 7 going: Sevennn.
• Your turn: Get 7 going. (Signal.) *Sevennn.*
• Get 9 going. (Signal.) *Niiine.*
• Get 4 going. (Signal.) *Fouuur.*
c. My turn to get 14 going: Fourteeen.
• Your turn: Get 14 going. (Signal.) *Fourteeen.*
• Get 16 going. (Signal.) *Sixteeen.*
• Get 21 going. (Signal.) *Twenty-wuuun.*
(Repeat steps a through c until firm.)

d. I'm going to start with 4 and count to 10. What number will I start with? (Signal.) *4.*
• I'll get 4 going and count to 10. Fouuur. (Tap 6.) 5, 6, 7, 8, 9, 10.
e. Your turn to start with 4 and count to 10.
• Get 4 going. *Fouuur.* Count. (Tap 6.) *5, 6, 7, 8, 9, 10.*
(Repeat steps d and e until firm.)
f. Your turn to start with 6 and count to 10.
• Get 6 going. *Siiix.* Count. (Tap 4.) *7, 8, 9, 10.*
(Repeat step f until firm.)
g. Your turn to start with 8 and count to 10.
• Get 8 going. *Eieieight.* Count. (Tap 2.) *9, 10.*
(Repeat step g until firm.)

Lesson 5, Exercise 6

Teaching Note: In steps A through C, you direct students to get different numbers going. In the following steps, students count from one specified number to another.

In step D, you model counting from 4 to 10. Then students count from 4 to 10.

In the following steps, students count from 6 to 10 and 8 to 10 without a model. It's probably a good idea to model step D two times so students see how long they hold to get 4 going.

Make sure that you use the same timing when you direct the students to start with 4 and count to 10. You say "Get 4 going." Make sure they hold "fouuur" the way you modeled it, then say "Count."

Students practice counting-on in the following lessons.

Adding Tens

On Lesson 9, students learn an important extension of counting by tens, which is a series of numbers that add 10.

For example: 23, 33, 43, 53.

Here's the introduction from Lesson 9:

EXERCISE 4: ADDING TENS   [REMEDY]

a. My turn to plus tens. I'll start with 21 and plus tens to 51.
• What number will I start with? (Signal.) *21.*
• What number will I plus tens to? (Signal.) *51.*
b. Here I go: Twenty-wuuun, 31, 41, 51. Your turn: Start with 21 and plus tens to 51.
• Get 21 going. *Twenty-wuuun.* Plus tens. (Tap 3.) *31, 41, 51.*
(Repeat step b until firm.)
c. My turn to start with 22 and plus tens to 52.
• What number will I start with? (Signal.) *22.*
• What number will I plus tens to? (Signal.) *52.*
d. Here I go: Twenty-twooo, 32, 42, 52. Your turn: Start with 22 and plus tens to 52.
• Get 22 going. *Twenty-twooo.* Plus tens. (Tap 3.) *32, 42, 52.*
(Repeat step d until firm.)
e. My turn to start with 25 and plus tens to 55.
• What number will I start with? (Signal.) *25.*
• What number will I plus tens to? (Signal.) *55.*
f. Here I go: Twenty-fiiive, 35, 45, 55. Your turn: Start with 25 and plus tens to 55.
• Get 25 going. *Twenty-fiiive.* Plus tens. (Tap 3.) *35, 45, 55.*
(Repeat step f until firm.)
g. My turn to start with 26 and plus tens to 56.
• What number will I start with? (Signal.) *26.*
• What number will I plus tens to? (Signal.) *56.*
h. Here I go: Twenty-siiix, 36, 46, 56. Your turn: Start with 26 and plus tens to 56.
• Get 26 going. *Twenty-siiix.* Plus tens. (Tap 3.) *36, 46, 56.*
(Repeat step h until firm.)
i. Your turn again: Start with 25 and plus tens to 55.
• What number will you start with? (Signal.) *25.*
• What number will you plus tens to? (Signal.) *55.*
• Get 25 going. *Twenty-fiiive.* Plus tens. (Tap 3.) *35, 45, 55.*
(Repeat step i until firm.)

Lesson 9, Exercise 4

This counting is important in *CMC Level B* because students will later apply it when they add ten to two-digit numbers (37 + 10) and 3-digit numbers (352 + 10). They will be able to work these problems as mental math, not computation.

A further application of adding ten is counting money.

Students count to 27 and then add 10 for the dime. (See page 82, Money.)

Counting by Hundreds

Counting hundreds begins on Lesson 21. Students learn to count by hundreds to 900. Students extend counting by hundreds to a thousand on Lesson 38.

Here's the exercise from Lesson 21:

EXERCISE 6: COUNTING BY HUNDREDS `REMEDY`

a. My turn to count by hundreds. One hundred, 200, 300, 400, 500, 600, 700, 800, 900.

b. Your turn: Count by hundreds to 900. Get ready. Count. **(Tap 9.)** *100, 200, 300, 400, 500, 600, 700, 800, 900.*
(Repeat step b until firm.)

c. What comes after 300, when you count by hundreds? **(Signal.)** *400.*

• What comes after 700, when you count by hundreds? **(Signal.)** *800.*

• What comes after 500, when you count by hundreds? **(Signal.)** *600.*

Lesson 21, Exercise 6

On later lessons, students add 100 to other hundreds numbers, for instance: "What's 421 plus 100?"

Counting Backward by Hundreds

Counting backward by tens from 100 begins on Lesson 31. Counting backward by hundreds begins on Lesson 33.

Here's part of the exercise from Lesson 33. Students count backward by both tens and hundreds.

---

c. Now you're going to count by tens to 100. What are you going to count by? **(Signal.)** *Tens.*

• Count by tens to 100. **(Tap.)** *10, 20, 30, 40, 50, 60, 70, 80, 90, 100.*

d. My turn to start with 100 and count backward by tens. One huuundred, 90, 80, 70, 60, 50, 40, 30, 20, 10.

• Your turn: Start with 100 and count backward by tens. Get 100 going. *One huuundred.* Count backward. **(Tap.)** *90, 80, 70, 60, 50, 40, 30, 20, 10.*
(Repeat step d until firm.)

e. My turn to start with 900 and count backward by hundreds. Nine huuundred, 800, 700, 600, 500, 400, 300, 200, 100.

• Your turn: Start with 900 and count backward by hundreds. Get 900 going. *Nine huuundred.* Count backward. **(Tap.)** *800, 700, 600, 500, 400, 300, 200, 100.*
(Repeat step e until firm.)

---

from Lesson 33, Exercise 1

Counting by 5s

Counting by fives begins on Lesson 48.

The numbers are initially displayed so they show a repeated pattern. The first number ends in 5. The second is the following tens number.

EXERCISE 4: COUNTING BY FIVES

a. (Display:)   [48:4A]

5	10
15	20
25	30
35	40
45	50

(Point to **10**.) The numbers in this column are numbers for counting by 10.
- Read the numbers. Get ready. (Touch tens numbers.) *10, 20, 30, 40, 50.*

b. (Point to **5**.) If I say all of the numbers, I count by 5. My turn to count by fives to 20: 5, 10, 15, 20.
- Your turn: Say the first four numbers with me. Get ready. (Touch numbers as you and children say:) *5, 10, 15, 20.*
(Repeat step b until firm.)

c. (Point to **5**.) Your turn: Say those numbers. Get ready. (Touch.) *5, 10, 15, 20.*
(Repeat until firm.)
- Say the first four numbers without looking. Get ready. (Signal.) *5, 10, 15, 20.*

d. (Point to **25**.) Say the next two numbers with me. Get ready. (Touch numbers.) *25, 30.*

e. (Point to **25**.) Your turn: Say those numbers. Get ready. (Touch numbers.) *25, 30.*

f. (Point to **5**.) Say the numbers for counting by fives to 30. Get ready. (Touch numbers.) *5, 10, 15, 20, 25, 30.*

g. (Do not show display.) My turn to count by fives to 30: 5, 10, 15, 20, 25, 30.
Your turn to count by fives to 30: Get ready. (Tap.) *5, 10, 15, 20, 25, 30.*
(Repeat step g until firm.)

Lesson 48, Exercise 4

Teaching Note: In step B, you model counting. For this pattern, touch and say the two numbers in a row, then pause before presenting the next two numbers.

     5, 10 (pause) 15, 20 (pause).

This pacing reinforces the notion that a number that ends in 5 is followed by a tens number.

On the following lessons, students assume more responsibility in generating the numbers. After students count by 5 to 50 while looking at the display, numbers are systematically removed from the display until the students count by fives without seeing the numbers.

Later students start in the middle of the sequence and count to 50.

The sequence is then extended beyond 50 to 100. After mastering counting from a specified fives number to another fives number, students count by fives starting with 3-digit numbers. ("Start with 365 and count by fives to 390.")

Students also apply counting by fives to counting money. For example:

Students count by fives to determine there are 30 dollars; then they count by fives for the nickels; and by ones for the penny.

Counting by 2s

Counting by twos is introduced on Lesson 93. The instruction is similar to that for fives, but the pattern for twos is greatly different from that for fives.

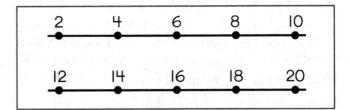

The pattern repeats after five numbers.

The work on twos initially focuses on only the top row.

Here's the exercise from Lesson 93:

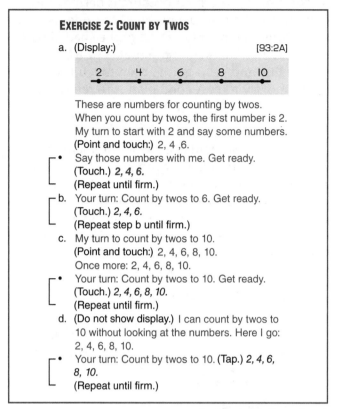

EXERCISE 2: COUNT BY TWOS

a. (Display:) [93:2A]

These are numbers for counting by twos.
When you count by twos, the first number is 2.
My turn to start with 2 and say some numbers.
(Point and touch:) 2, 4 ,6.
• Say those numbers with me. Get ready.
(Touch.) *2, 4, 6.*
(Repeat until firm.)

b. Your turn: Count by twos to 6. Get ready.
(Touch.) *2, 4, 6.*
(Repeat step b until firm.)

c. My turn to count by twos to 10.
(Point and touch:) 2, 4, 6, 8, 10.
Once more: 2, 4, 6, 8, 10.
• Your turn: Count by twos to 10. Get ready.
(Touch.) *2, 4, 6, 8, 10.*
(Repeat until firm.)

d. (Do not show display.) I can count by twos to
10 without looking at the numbers. Here I go:
2, 4, 6, 8, 10.
• Your turn: Count by twos to 10. (Tap.) *2, 4, 6,
8, 10.*
(Repeat until firm.)

Lesson 93, Exercise 2

On Lesson 97, students learn that the numbers from 12 through 20 have the same ones digit as numbers 2 through 10. After students are practiced in counting by twos, they extend the pattern to numbers beyond 20. By Lesson 110, students are able to count by twos from 50 to 60, from 20-40, or from 180 to 190.

COUNT-ON STRATEGY

Starting on Lesson 49, students use a count-on strategy to figure out the number of objects in a display. Students write an addition problem that tells about the objects in two groups. They start with the number for the first group. Then they count on to figure out the total number of objects in both groups.

Here's a problem of this type:

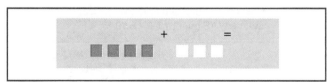

Students write the numbers for the shaded boxes and the unshaded boxes. Then they get 4 going and count-on for the unshaded boxes. *Four, 5, 6, 7.* Students write the answer.

On Lesson 57, students work with larger numbers. The first group in each problem is not counted, but the number above it tells how many are in the group.

Here's the exercise from Lesson 57:

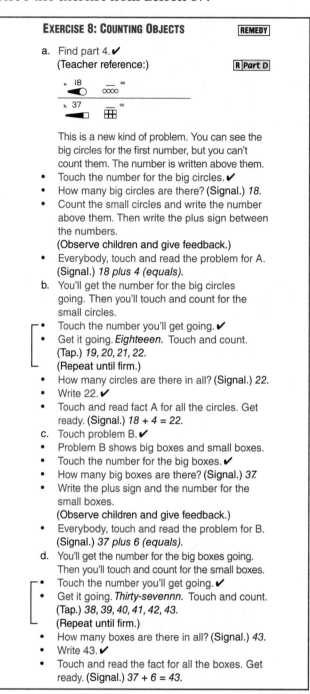

EXERCISE 8: COUNTING OBJECTS [REMEDY]

a. Find part 4. ✔
(Teacher reference:) [R Part D]

This is a new kind of problem. You can see the big circles for the first number, but you can't count them. The number is written above them.
• Touch the number for the big circles. ✔
• How many big circles are there? (Signal.) *18.*
• Count the small circles and write the number above them. Then write the plus sign between the numbers.
(Observe children and give feedback.)
• Everybody, touch and read the problem for A.
(Signal.) *18 plus 4 (equals).*
b. You'll get the number for the big circles going. Then you'll touch and count for the small circles.
• Touch the number you'll get going. ✔
• Get it going. *Eighteeen.* Touch and count.
(Tap.) *19, 20, 21, 22.*
(Repeat until firm.)
• How many circles are there in all? (Signal.) *22.*
• Write 22. ✔
• Touch and read fact A for all the circles. Get ready. (Signal.) *18 + 4 = 22.*
c. Touch problem B. ✔
• Problem B shows big boxes and small boxes.
• Touch the number for the big boxes. ✔
• How many big boxes are there? (Signal.) *37.*
• Write the plus sign and the number for the small boxes.
(Observe children and give feedback.)
• Everybody, touch and read the problem for B.
(Signal.) *37 plus 6 (equals).*
d. You'll get the number for the big boxes going. Then you'll touch and count for the small boxes.
• Touch the number you'll get going. ✔
• Get it going. *Thirty-sevennn.* Touch and count.
(Tap.) *38, 39, 40, 41, 42, 43.*
(Repeat until firm.)
• How many boxes are there in all? (Signal.) *43.*
• Write 43. ✔
• Touch and read the fact for all the boxes. Get ready. (Signal.) *37 + 6 = 43.*

Lesson 57, Exercise 8

Teaching Note: If students have trouble touching and counting accurately, tell them that you'll do the counting. They'll do the touching. Following a successful example, return to the original exercise:

"Your turn to do the touching **and** counting."

Starting on Lesson 60, students work with pictures of rulers without numbers.

The purpose of this work is to make sure students are firm in their understanding of intervals on a ruler. Specifically, the first mark on the ruler is not the beginning of the first unit but the end of the first unit. The next mark shows two completed intervals. And so forth.

On Lesson 62, students work with rulers that have numbers for the shaded part and the unshaded part.

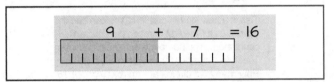

from Lesson 62, Exercise 1

Students confirm that the numbers for the equation are correct. They touch the end of the shaded part. Get the number for that part going. Then touch and count on for the marks on the unshaded part.

Here's part of the exercise from Lesson 63:

---

c. Find part 4 on worksheet 63. ✔
(Teacher reference:) [R] Part A

- Touch ruler A. ✔
  It shows centimeters.
- Touch the number of centimeters for the shaded part. ✔
- How many centimeters? (Signal.) *3.*
- Touch the number of centimeters for the unshaded part. ✔
- How many centimeters? (Signal.) *6.*
d. You're going to figure out how many centimeters there are for both parts.
- Touch the end of the shaded part. ✔
- What number will you get going? (Signal.) *3.*
- Get it going. *Threee.* Touch and count the lines. (Tap.) *4, 5, 6, 7, 8, 9.*
  (Repeat until firm.)
- How many centimeters are there for both parts? (Signal.) *9.*
- Write 9 to complete the fact for both parts. ✔
- Everybody, touch and read the fact for ruler A. (Signal.) *3 + 6 = 9.*

---

from Lesson 63, Exercise 7

Teaching Note: The procedure is designed to minimize the possible problem of students touching the line at the end of the shaded part and saying four. Make sure that they touch the third line and get 3 going, then touch the next line and say four. A good plan is to repeat this counting at least one time.

By Lesson 68, students also work problems that do not have a number for the unshaded part of the ruler. Also, these problems do not show the intervals for the shaded part, just the number. Students first count for the unshaded part and write the number. Then start with the number for the shaded part and count on to figure out the length of the whole ruler.

Here's part of the exercise from Lesson 68:

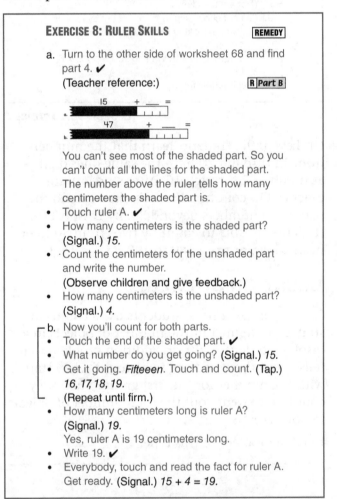

from Lesson 68, Exercise 8

On Lesson 70, students count on backward to figure out the missing number for the shaded part of the ruler.

Connecting Math Concepts

Here's part of the exercise from Lesson 70:

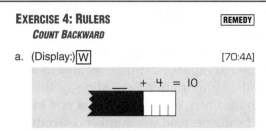

EXERCISE 4: RULERS   REMEDY
COUNT BACKWARD

a. (Display:) W   [70:4A]

$$\underline{\quad} + 4 = 10$$

This ruler is supposed to show inches. It has a shaded part (touch) and an unshaded part (touch). You're going to figure out how many inches are shaded.
- (Point to **4**.) How many inches is the unshaded part? (Touch.) *4.*
- (Point to **10**.) How many inches long is the ruler? (Touch.) *10.*

b. We can figure out how many inches are shaded by counting backward from the end of the ruler to the shaded part.
- (Point to the end of the ruler.) My turn: (touch the end of the ruler) *Tennn,* (touch lines) *9, 8, 7, 6.*

c. (Point to the end of the ruler.) Your turn to get ten going and count backward to the shaded part.
- (Touch the end of the ruler.) Get it going. *Tennn.* Count backward. (Touch lines.) *9, 8, 7, 6.* (Repeat until firm.)
- Everybody, how many inches is the shaded part? (Tap.) *6.*
(Add to show:)   [70:4B]

$$\underline{6} + 4 = 10$$

- (Point to **6**.) Read the equation. Get ready. (Touch.) *6 + 4 = 10.*
- (Point to **6**.) How many inches is the shaded part? (Touch.) *6.*
- (Point to **4**.) How many inches is the unshaded part? (Touch.) *4.*
- (Point to **10**.) How many inches is the whole ruler? (Touch.) *10.*

from Lesson 70, Exercise 4

Teaching Note: The bracketed segment in step C shows the critical part of the counting. You touch the end of the ruler and say: "Get it going." Students hold *tennn* as long as you touch the end of the ruler.

When you touch the intervals of the unshaded part, students count: 9, 8, 7, 6. Repeat this part if necessary. Then students tell you how many inches the shaded part is.

On later lessons, students apply the count-on backward strategy to larger numbers.

Here are the examples from Lesson 74:

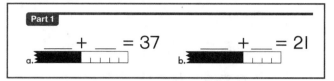

Part 1

a. $$\underline{\quad} + \underline{\quad} = 37$$ b. $$\underline{\quad} + \underline{\quad} = 21$$

from Lesson 74, Exercise 6

The later independent work presents mixed sets of problems. Some have objects, not rulers. Some rulers show the number of units for the shaded part. Others show the total units.

Here's a problem set from Lesson 90:

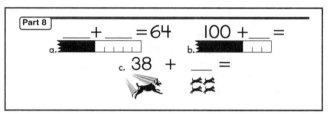

Part 8

a. $$\underline{\quad} + \underline{\quad} = 64$$ b. $$100 + \underline{\quad} =$$

c. $$38 + \underline{\quad} =$$

from Workbook Lesson 90, Part 8

Symbol Identification

The primary purpose of the Symbol Identification exercises is to provide students with fluency in reading and writing numbers and signs. In the early lessons, students review reading and writing single-digit numbers, teens, decade numbers (40, 70, etc.), and two-part numbers (46, 72, etc.). The first new numbers introduced are hundreds numbers beyond 199.

READING NUMBERS

Students learn to identify numbers 1–199 in *CMC Level A.* Lessons 1 through 20 review these numbers with daily exercises.

On Lesson 21, students learn about digits. They answer questions about the ones digit and the tens digit.

- 14. Read the number.
- How many digits does 14 have?
- What's the ones digit?
- What's the tens digit?

The first new numbers taught in *CMC Level B* are 3-digit numbers that are more than 199.

Here's the exercise from Lesson 21:

EXERCISE 3: SYMBOL IDENTIFICATION
3-DIGIT NUMBERS

a. (Display:)   [21:3A]

156   104
356   504   7<u>13</u>
856   704

- (Point to <u>156</u>.) What's the underlined part? (Touch.) *56.*
- What's the whole number? (Touch.) *156.*
- (Point to **356**.) This number is 356. What number? (Touch.) *356.*
- (Point to **856**.) Look at this number. What number? (Touch.) *856.*
- (Repeat until firm.)
b. (Point to **104**.) What do you say for the underlined part of this number? (Touch.) *4.*
- What's the whole number? (Touch.) *104.*
- (Point to **504**.) This number is 504. What number. (Touch.) *504.*
- (Point to **704**.) Look at this number. What number. (Touch.) *704.*
c. (Point to **713**.) What do you say for the underlined part of this number? (Touch.) *13.*
- What's the whole number? (Touch.) *713.*
d. Let's read those numbers again.
- (Point to **156**.) What number? (Touch.) *156.*
- (Repeat for remaining numbers.)
- (Repeat step d until firm.)

Lesson 21, Exercise 3

Teaching Note: The first number in each column has the tens and ones underlined. Students first read the underlined part, then the whole number. For the other numbers in the column, the only difference is the hundreds number. You model how to read the second number in the column. For the third number, you simply ask, "What number?"

Students quickly learn numbers beyond 199. The program provides daily practice with isolated numbers and later practice with addition and subtraction problems that have 3-digit numbers.

WRITING NUMBERS

Exercises that involve writing numbers begin on Lesson 1.

On Lesson 7, students learn the rule that if you say most 2-digit numbers, you know how to write them. This rule is important for guarding against reversals of teen numbers and numbers that end in 1 (81, 41, etc.). If students read the number correctly, they say the digits in the correct order "Eighty-one."

Here's the exercise from Lesson 7 that introduces the rule:

EXERCISE 7: TWO-PART NUMBERS
40s & 60s   REMEDY

a. When you say some numbers with two parts, you know how to write the numbers. My turn: 64. What's the first part of 64? 6. What's the other part? 4.
- Your turn: 64. What's the first part? (Signal.) *6.* What's the other part? (Signal.) *4.*
b. New number: 62. What number? (Signal.) *62.* What's the first part? (Signal.) *6.* What's the other part? (Signal.) *2.*
c. New number: 63. What number? (Signal.) *63.* What's the first part? (Signal.) *6.* What's the other part? (Signal.) *3.*
d. New number: 61. What number? (Signal.) *61.* What's the first part? (Signal.) *6.* What's the other part? (Signal.) *1.*
(Repeat steps b through d that were not firm.)
e. My turn: 41. What's the first part of 41? 4. What's the other part? 1.
- Your turn: 41. What's the first part? (Signal.) *4.* What's the other part? (Signal.) *1.*
f. New number: 49. What number? (Signal.) *49.* What's the first part? (Signal.) *4.* What's the other part? (Signal.) *9.*
g. New number: 46. What number? (Signal.) *46.* What's the first part? (Signal.) *4.* What's the other part? (Signal.) *6.*
h. New number: 43. What number? (Signal.) *43.* What's the first part? (Signal.) *4.* What's the other part? (Signal.) *3.*
(Repeat steps f through h that were not firm.)

Lesson 7, Exercise 7

Teaching Note: The exercise refers to parts because the term *digits* has not been introduced yet. Digits are introduced on Lesson 21.

Between Lessons 21 and 31, students write 2-digit numbers. Sets of examples typically have numbers that some students reverse, such as 15 and 51.

For some writing exercises, students work from dictation. For others, they identify digits and copy numbers that are written. The goal of the track is for students to become proficient in identifying the hundreds, tens, and ones digits of different types of numerals and proficient in writing numbers from dictation. One criterion for writing is that they write the digits in appropriate columns. This work is essential to proper computation, which shows the hundreds, tens, and ones lined up in columns.

On Lesson 31, students write 3-digit numbers. The early sets are made up of numbers that are not teens and that do not have zeros.

Here's part of the exercise from Lesson 31:

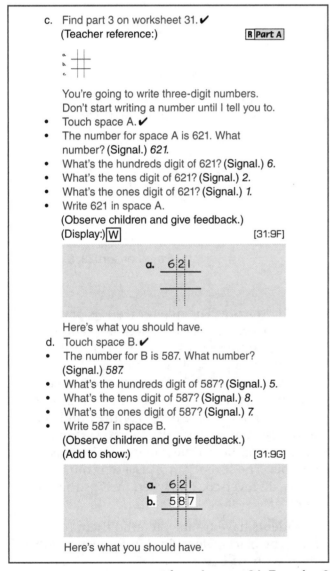

from Lesson 31, Exercise 9

The grid prompts students to write each digit in the appropriate column. In later lessons, students write sets that include one digit, 2-digit, and 3-digit numbers.

Students will continue to work the variations of grids throughout *CMC Level B*.

Place Value

Place Value is developed in two directions: the relationship of the name to the symbols (forty-two and 42); the relationship of the names to addition facts (40 + 2 = 42).

The first direction is developed as part of symbol writing. The relationship of names to addition facts begins on Lesson 16.

Here's the exercise:

EXERCISE 1: PLACE-VALUE ADDITION REMEDY

a. I'll say the equation for 70 plus 4. 70 plus 4 equals 74.
• Say the equation for 70 plus 4. (Signal.) *70 plus 4 equals 74.*
(Repeat step a until firm.)
b. Say the equation for 70 plus 9. (Signal.) *70 plus 9 equals 79.*
• Say the equation for 40 plus 9. (Signal.) *40 plus 9 equals 49.*
• (Repeat for 40 + 1, 90 + 6, 30 + 5, 50 + 3.)
(Repeat step b until firm.)

INDIVIDUAL TURNS

Now I'll call on individuals.
(Call on individual students to perform one or two of the following tasks.)

• Say the equation for 40 plus 1. (Call on a student.) *40 + 1 = 41.*
• Say the equation for 90 plus 6. (Call on a student.) *90 + 6 = 96.*
• Say the equation for 30 plus 5. (Call on a student.) *30 + 5 = 35.*

Lesson 16, Exercise 1

Teaching Note: Establish a rhythm for saying the place-value addition. "Thirty plus five equals *thirtyyy*-five."

This exercise presents individual turns. Make sure students respond very well on the group tasks before you present Individual Turns. So you may want to repeat step B before going on in the exercise.

Note also that students should be able to respond without hesitation when you call on them. Make sure that they use the same rhythm pattern that you use when modeling the addition.

Starting on Lesson 19, students complete place-value addition equations. The number is shown in the equation.

$$\underline{\qquad} + \underline{\qquad} = 37$$

Students say the place-value addition for 37; then write the missing numbers:

$$\underline{\ 30\ } + \underline{\ 7\ } = 37$$

Here's the part of the exercise in which students complete the equations:

---

e. Find part 2 on your worksheet. ✔
(Teacher reference:)   R Part D

  a. __ + __ = 37
  b. __ + __ = 81
  c. __ + __ = 52
  d. __ + __ = 94

You're going to say the place-value addition for each problem.
- Touch the number in problem A. ✔
- What number? (Signal.) *37.*
- Say the place-value addition for 37. (Signal.) *30 + 7 = 37.*
(Repeat until firm.)
- Touch where you'll write 30. ✔
- Touch where you'll write 7. ✔
f. Touch the number in problem B. ✔
- What number? (Signal.) *81.*
- Say the place-value addition for 81. (Signal.) *80 + 1 = 81.*
g. Touch the number in problem C. ✔
- What number? (Signal.) *52.*
- Say the place-value addition for 52. (Signal.) *50 + 2 = 52.*
h. Touch D. ✔
- What number? (Signal.) *94.*
- Say the place-value addition for 94. (Signal.) *90 + 4 = 94.*
i. Complete the place-value equations in part 2. (Observe children and give feedback.)
j. Check your work. You'll touch and read each place-value addition equation.
- Equation A. (Signal.) *30 + 7 = 37.*
- (Repeat for:) B, *80 + 1 = 81;* C, *50 + 2 = 52;* D, *90 + 4 = 94.*

---

from Lesson 19, Exercise 9

Writing equations for teen numbers begins on Lesson 22.

Here's part of the exercise:

---

b. (Display:)   [22:10A]

$$15 \qquad 18$$
$$12 \qquad 11$$

- (Point to **15**.) What number is this? (Touch.) *15.* My turn to say the place-value addition for 15: 10 plus 5 equals 15.
- Say the place-value addition for 15. (Signal.) *10 + 5 = 15.*
c. (Point to **18**.) What number is this? (Touch.) *18.* My turn to say the place-value addition for 18: 10 plus 8 equals 18.
- Say the place-value addition for 18. (Signal.) *10 + 8 = 18.*
d. (Point to **12**.) What number is this? (Touch.) *12.*
- Say the place-value addition for 12. (Signal.) *10 + 2 = 12.*
e. (Point to **11**.) What number is this? (Touch.) *11.*
- Say the place-value addition for 11. (Signal.) *10 + 1 = 11.*
- (Point to **15**.) Say the place-value addition for 15 again. (Signal.) *10 + 5 = 15.*
- (Point to **18**.) Say the place-value addition for 18 again. (Signal.) *10 + 8 = 18.*
(Repeat step e until firm.)
Remember, the place-value addition for teen numbers starts with 10.

---

from Lesson 22, Exercise 10

Teaching Note: The rule for teens is presented at the end of step E: The place-value addition for teen numbers starts with 10. Remember this rule. You probably will have many opportunities to remind students of it. The reason is that the digit named first in most teen numbers refers to the ones. For the numbers students have worked with before Lesson 22, the first number in the name tells the tens number. The seventy in 71 is a tens number. Teens are reversed. The seven in 17 is the ones number.

If students have trouble, remind them of the rule: The place-value addition for teen numbers starts with 10.

Place-value addition for decade numbers starts on Lesson 24. These numbers are different because they don't name the ones number.

Three-digit numbers are introduced on Lesson 29. The format of the introduction is that you say the place-value addition; the students say the number it equals.

Here's the exercise:

EXERCISE 5: PLACE-VALUE ADDITION
3-DIGIT NUMBERS   `REMEDY`

a. Listen: 400 plus 90 plus 3 equals 493.
• What's 400 plus 90 plus 3? (Signal.) *493.*
b. Listen: What's 100 plus 70 plus 4? (Signal.) *174.*
• Listen: What's 600 plus 20 plus 7? (Signal.) *627.*
• Listen: What's 500 plus 30 plus 8? (Signal.) *538.*
• Listen: What's 200 plus 80 plus 3? (Signal.) *283.*
  (Repeat step b until firm.)
c. Listen to a three-digit place-value equation: 400 plus 20 plus 8 equals 428. Say it with me. Get ready. (Signal.) **400 plus 20 plus 8 equals 428.**
• Your turn: Say the place-value equation for 428. Get ready. (Signal.) *400 + 20 + 8 = 428.*
  (Repeat step c until firm.)

Lesson 29, Exercise 5

On later lessons, you say the number and students either say the place-value addition or write it.

Here's the set of problems from Lesson 37:

`Part 1`

a. ____ + ___ + __ = 782

b. ____ + ___ + __ = 230

c. ____ + ___ + __ = 193

d. ____ + ___ + __ = 850

Workbook Lesson 37, Part 1

After Lesson 50, students write place-value addition in columns.

Number Families

Students who have gone through *CMC Level A* have learned count-on strategies for working addition and subtraction problems, including problems that traditionally require carrying.

For example: $44 + 26 =$

Students solve the problem by making Ts and lines for 26. Each T represents ten; each line represents one.

$$44 + 26 =$$
$$\textbf{TTIIIIII}$$

Students then start with 44 and count first for Ts and then for lines: *54, 64, 65, 66, 67, 68, 69, 70.*

The operations that students learn for *CMC Level B* don't involve counting on. Rather, students learn facts. The primary vehicle for this learning is "number families." Each family consists of three numbers that always go together to create addition and subtraction facts.

NUMBER FAMILY STRATEGY

Each number family has two small numbers and a big number.

For example: $\xrightarrow[\hspace{2cm}]{5 \qquad 2} 7$.

The small numbers in this family are 5 and 2. The big number is 7.

The family generates two addition facts and two subtraction facts.

Addition facts begin with a small number:

$$5 + 2 = 7$$
$$2 + 5 = 7$$

Subtraction facts start with the big number:

$$7 - 2 = 5$$
$$7 - 5 = 2$$

The logic of number families is that if you know the three numbers that go together in a family, you can figure out the missing number in a family that shows only two numbers.

$$\xrightarrow[\hspace{2cm}]{5 \qquad 2} \_$$
$$\xrightarrow[\hspace{2cm}]{5 \qquad} 7$$

There are two rules for finding the missing number:

1. If the big number is missing, you add to find it. (For the top example, students work the problem $5 + 2 =$.)

2. If a small number is missing, you subtract to find it. (For the bottom example, students work the problem: $7 - 5 =$.)

A variation is a number family that does not involve familiar facts.

$$\xrightarrow[\hspace{2cm}]{\_ \qquad 35} 47$$

Students work the problem:

$$\begin{array}{r} 47 \\ - 35 \\ \hline \end{array}$$

Number families are introduced on Lesson 16. The families that students learn about in this lesson generate plus-1 and minus-1 facts students know.

In this exercise, students learn to identify the small numbers and big number.

Here is the exercise:

EXERCISE 8: NUMBER FAMILIES

a. (Display:)   [16:8A]

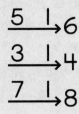

These are **number families.**
* What are they? (Signal.) *Number families.*
  If you learn number families, you don't have to count to figure out the answer to problems that plus or minus.
  Remember: If you know the three numbers in a family, you don't have to count to work plus or minus problems with these numbers.
b. Each number family is on an arrow. Each number family has three numbers.
* How many numbers does each number family have? (Signal.) *Three.*
* (Point to 5→6.) How many numbers are in this family? (Touch.) *Three.*
* (Point to 3→4.) How many numbers are in this family? (Touch.) *Three.*
* (Point to 7→8.) How many numbers are in this family? (Touch.) *Three.*
c. Two numbers are small numbers.
* How many are small numbers? (Signal.) *Two.*
* (Point to **5** and **1.**) The small numbers in this family are 5 and 1. What are the small numbers in this family? (Touch.) *5 and 1.*
* (Point to **6.**) The big number is 6. What's the big number? (Touch.) *6.*
d. (Point to 5→6.) Again, what are the small numbers? (Signal.) *5 and 1.*
* What's the big number? (Signal.) *6.*
  (Repeat step d until firm.)
e. (Point to **3** and **1.**) The small numbers in this family are 3 and 1. What are the small numbers? (Touch.) *3 and 1.*
* (Point to **4.**) This is the big number. What's the big number? (Touch.) *4.*
f. (Point to 3→4.) What are the small numbers? (Signal.) *3 and 1.*
  What's the big number? (Signal.) *4.*
* (Point to 7→8.) What are the small numbers in this family? (Touch.) *7 and 1.*
  What's the big number? (Touch.) *8.*
g. Let's do those again.
* (Point to 5→6.) What are the small numbers in this family? (Signal.) *5 and 1.*
  What's the big number? (Signal.) *6.*
* (Point to 3→4.) What are the small numbers in this family? (Signal.) *3 and 1.*
  What's the big number? (Signal.) *4.*
* (Point to 7→8.) What are the small numbers in this family? (Signal.) *7 and 1.*
  What's the big number? (Signal.) *8.*
  (Repeat step g until firm.)

Lesson 16, Exercise 8

Teaching Note: Make sure that students respond well in step D. If they do, they will have a good foundation for what occurs on the following lessons.

On Lesson 20, students generate two addition facts for families. Note that the script uses the word *plus*, not *add*. The reason for this convention is that students say "plus," (not "add,") when they read problems.

Here's part of the exercise:

---

f. (Point to 5→.) Say the plus problem that starts with the first small number. (Touch.) *5 plus 1.*
* Say the plus problem that starts with 1. (Touch.) *1 plus 5.*
* What does 1 plus 5 equal? (Signal.) *6.*
  That's the big number in the family.
  (Add to show:)   [20:4B]

  $$\underline{5} \xrightarrow{\ \ 1\ \ } \underline{6}$$

* (Point to 5→.) Say the **fact** that starts with 5. (Touch numbers.) *5 + 1 = 6.*
* Say the fact that starts with 1. (Touch.) *1 + 5 = 6.*
  (Repeat until firm.)
g. (Point to 7→.) Say the plus problem that starts with 7. (Touch.) *7 plus 1.*
* Say the plus problem that starts with 1. (Touch.) *1 plus 7.*
* What does 1 plus 7 equal? (Signal.) *8.*
  (Add **8** to show.)   [20:4C]
* (Point to 7→.) Say the fact that starts with 7. (Touch.) *7 + 1 = 8.*
* Say the fact that starts with 1. (Touch.) *1 + 7 = 8.*

---

from Lesson 20, Exercise 4

Teaching Note: In the first part of the exercise (not shown), students identified the small numbers and said problems for figuring out the missing big numbers.

For each number family, students say the problem and identify the missing big number. Next, they say two facts: the fact that starts with the first small number, and the fact that starts with the other small number.

Repeat any steps that have weak responses.

On Lesson 24, students learn that minus facts start with the big number of the family. One fact goes straight backward along the arrow. Note that the script uses the word *minus*, not *subtract*. The reason for this convention is that students say "minus" (not subtract) when they read problems.

Here's part of the exercise:

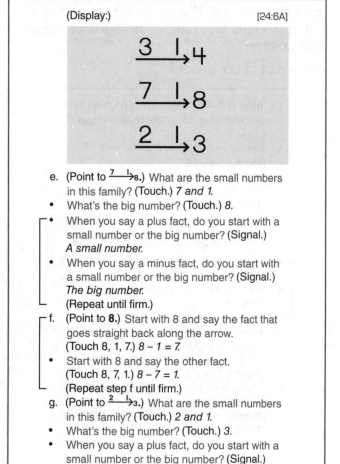

(Display:)   [24:6A]

e. (Point to 7 ↘8.) What are the small numbers in this family? (Touch.) *7 and 1.*
   • What's the big number? (Touch.) *8.*
   • When you say a plus fact, do you start with a small number or the big number? (Signal.) *A small number.*
   • When you say a minus fact, do you start with a small number or the big number? (Signal.) *The big number.*
     (Repeat until firm.)
f. (Point to **8.**) Start with 8 and say the fact that goes straight back along the arrow. (Touch 8, 1, 7.) *8 – 1 = 7.*
   • Start with 8 and say the other fact. (Touch 8, 7, 1.) *8 – 7 = 1.*
     (Repeat step f until firm.)
g. (Point to 2 ↘3.) What are the small numbers in this family? (Touch.) *2 and 1.*
   • What's the big number? (Touch.) *3.*
   • When you say a plus fact, do you start with a small number or the big number? (Signal.) *A small number.*
   • When you say a minus fact, do you start with a small number or the big number? (Signal.) *The big number.*
h. (Point to **3.**) Start with 3 and say the fact that goes straight back along the arrow. (Touch.) *3 – 1 = 2.*
   • Start with 3 and say the other minus fact. (Touch 3, 2, 1.) *3 – 2 = 1.*
     (Repeat step h until firm.)

from Lesson 24, Exercise 6

Teaching Note: It is very important for students to remember the pair of questions that appears first for each family.

   • When you say a minus fact, do you start with a small number or the big number?

   • When you say a plus fact, do you start with a small number or the big number?

If students are not firm on these relationships, they will have trouble later when they create number problems for families that have a missing number.

On Lesson 23, students write two familiar addition facts from number families.

Here's part of the exercise:

EXERCISE 9: NUMBER FAMILIES
WRITING PLUS FACTS   REMEDY

a. Find part 3 on your worksheet. ✔
   (Teacher reference:)   R Part A

   a. 5 1 _   b. 7 1 _   c. 9 1 _
   _____ _____ _____
   _____ _____ _____

   The families don't show the big number. You're going to say two facts for each family. Later you'll write both addition facts for each family.
   • Touch the first small number in family A. ✔
   • What's the first small number? (Signal.) *5.*
   • Say the fact that starts with 5. (Signal.) *5 + 1 = 6.*
   • Say the fact that starts with the other small number. (Signal.) *1 + 5 = 6.*
b. Touch the top space below family A. ✔
   • That's where you'll write 5 plus 1 equals 6. What fact will you write in the top space? (Signal.) *5 + 1 = 6.*
   • Touch where you'll write 1 plus 5 equals 6. ✔
     (Repeat step b until firm.)
c. Touch the first small number in family B. ✔
   • What's the first small number? (Signal.) *7.*
   • Say the fact that starts with 7. (Signal.) *7 + 1 = 8.*
   • Say the fact that starts with the other small number. (Signal.) *1 + 7 = 8.*
   • Touch where you'll write 7 plus 1 equals 8. ✔
   • Touch where you'll write 1 plus 7 equals 8. ✔

from Lesson 23, Exercise 9

On Lesson 20, students are introduced to families that have a missing number. Students first identify whether the missing number is a small number or the big number. If a small number is missing, they start with the big number and minus. If the big number is missing, they start with a small number and plus.

By Lesson 37, students have generated problems for sets of families that have either a missing small number or a missing big number. On Lesson 37, students work with the first set that has both families with a missing small number and families with a missing big number.

Here's part of the exercise from Lesson 37:

EXERCISE 3: NUMBER FAMILIES [REMEDY]

a. Think about saying plus problems for a family.
- When you plus, do you start with a small number or the big number? (Signal.) *A small number.*
- When you minus, what do you start with? (Signal.) *The big number.*
(Repeat step a until firm.)

b. (Display:) W [37:3A]

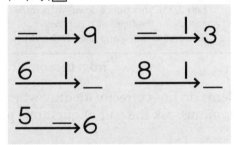

Some of these families have a missing small number. Others have a missing big number. Here's a rule: If a small number is missing, you minus.
- What do you do if a small number is missing? (Signal.) *Minus.*

c. (Point to ═╌▸9.) Is a small number missing? (Signal.) *Yes.*
- So do you minus? (Signal.) *Yes.*

d. (Point to 6 ╌▸_.) Is a small number missing? (Signal.) *No.*
- So do you minus? (Signal.) *No.*
So you plus.
- What do you do? (Signal.) *Plus.*

e. (Point to 5 ═▸6.) Is a small number missing? (Signal.) *Yes.*
- So do you minus? (Signal.) *Yes.*

f. (Point to ═╌▸3.) Is a small number missing? (Signal.) *Yes.*
- So do you minus? (Signal.) *Yes.*

g. (Point to 8 ╌▸_.) Is a small number missing? (Signal.) *No.*
- So do you minus? (Signal.) *No.*
- What do you do? (Signal.) *Plus.*

h. Now you're going to tell me the problem for each family.
- (Point to ═╌▸9.) Is a small number missing? (Signal.) *Yes.*
- So do you minus? (Signal.) *Yes.*
- Start with the big number and say the problem. (Touch.) *9 minus 1.*

i. (Point to 6 ╌▸_.) Is a small number missing? (Signal.) *No.*
- So do you minus? (Signal.) *No.*
- Start with the small number and say the problem. (Touch.) *6 plus 1.*

j. (Point to 5 ═▸6.) Is a small number missing? (Signal.) *Yes.*
- So do you minus? (Signal.) *Yes.*
- Start with the big number and say the problem. (Touch.) *6 minus 5.*

k. (Point to ═╌▸3.) Is a small number missing? (Signal.) *Yes.*
- So do you minus? (Signal.) *Yes.*
- Start with the big number and say the problem. (Touch.) *3 minus 1.*

l. (Point to 8 ╌▸_.) Is a small number missing? (Signal.) *No.*
- So do you minus? (Signal.) *No.*
- Start with the first small number and say the problem. (Touch.) *8 plus 1.*

m. This time, I'll say the problem for each family. You'll tell me the missing number.
- (Point to ═╌▸9.) Will I say a minus problem? (Signal.) *Yes.*
(Touch symbols.) 9 minus 1.
- What's the answer? (Signal.) *8.*
(Add to show:) [37:3B]

8 ╌▸9

n. (Point to 6 ╌▸_.) Will I say a minus problem? (Signal.) *No.*
(Touch symbols.) 6 plus 1.
- What's 6 plus 1? (Signal.) *7.*
(Add 7 to show.) [37:3C]

from Lesson 37, Exercise 3

The questions you ask in steps C through G are:

- Is a small number missing?
- So do you minus?

If the answer to the first question is no, the answer to the other question is no.

In this case, the big number is missing and you plus.

If the answer to both questions is yes, a small number is missing and you minus.

In steps H through L, students answer both questions and say the problem for finding the missing number.

The routines that you present in this exercise are designed to eliminate some of the confusion that students have in relating the missing number to the problem they work (or say). The wording that you use prompts students to start with the correct number.

- Start with the big number and say the problem.
- Start with the small number and say the problem.

Later, if students make mistakes, remind them how to say the problem for the missing number. Don't prompt them if they don't need it; but if they get stuck, ask them:

- Is a small number or the big number missing?
- So do you minus?
- Start with the _____ number and say the problem.

By Lesson 40, you introduce an abbreviated routine. First you go through the set of problems asking questions similar to the ones on Lesson 37. Then you go through the families the second time with less structure:

- Do you plus or minus for the missing number?
- Say the problem for the family.
- What's the missing number?

Here's the less-structured part of the exercise:

---

f. This time you'll say the problem for each family and tell me the missing number.
- Touch family A. ✔
- Do you minus or plus for the missing number? (Signal.) *Minus.*
- Say the problem for family A. Get ready. (Signal.) *9 minus 7.*
- What's the missing number? (Signal.) *2.*
g. Touch family B. ✔
- Do you minus or plus for the missing number? (Signal.) *Plus.*
- Say the problem for family B. Get ready. (Signal.) *6 plus 1.*
- What's the missing number? (Signal.) *7.*
h. Touch family C. ✔
- Do you minus or plus for the missing number? (Signal.) *Minus.*
- Say the problem for family C. Get ready. (Signal.) *8 minus 2.*
- What's the missing number? (Signal.) *6.*
i. Touch family D. ✔
- Do you minus or plus for the missing number? (Signal.) *Minus.*
- Say the problem for family D. Get ready. (Signal.) *7 minus 5.*
- What's the missing number? (Signal.) *2.*
j. Write the missing number for each family in part 2. Put your pencil down when you've completed the families in part 2.
(Observe children and give feedback.)

---

from Lesson 40, Exercise 6

If students do not correctly identify whether they plus or minus, ask the pair of questions practiced earlier:

- Is a small number missing?
- So do you minus?

After presenting the remaining steps in the exercise, return to any examples students missed and repeat them as written.

ADDITION/SUBTRACTION FACTS

Before the introduction of number families, students who begin at Lesson 1 review the facts that were taught to continuing students in *CMC Level A*: plus-1 facts, minus-1 facts, plus-10 facts, and plus-zero facts. Work with number families begins on Lesson 16. After students learn the features of number families, they learn facts through number families.

Number families for facts begin on Lesson 21 and continue throughout the level. Initially, students

use number families to review addition facts that involve plus 1, plus 2, plus 10, minus 1, and minus 2.

The facts that students have not learned are those that equal 1, 2, and 10: 5 minus 4, 5 minus 3, 16 minus 6. Work on these facts occurs while the familiar facts are reviewed in the lesson range of 21–49.

Starting on Lesson 50, students learn a new family:

$$\underline{5 \qquad 3}_{\rightarrow}8.$$

Here's the first part of the exercise from Lesson 51:

EXERCISE 7: NUMBER FAMILIES

a. (Display:)   [51:7A]

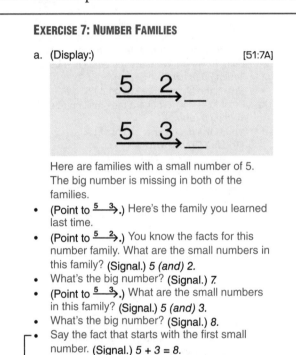

Here are families with a small number of 5. The big number is missing in both of the families.

- (Point to $\underline{5 \quad 3}_{\rightarrow}$.) Here's the family you learned last time.
- (Point to $\underline{5 \quad 2}_{\rightarrow}$.) You know the facts for this number family. What are the small numbers in this family? (Signal.) *5 (and) 2.*
- What's the big number? (Signal.) *7.*
- (Point to $\underline{5 \quad 3}_{\rightarrow}$.) What are the small numbers in this family? (Signal.) *5 (and) 3.*
- What's the big number? (Signal.) *8.*
- Say the fact that starts with the first small number. (Signal.) *5 + 3 = 8.*
- Say the other plus fact. (Signal.) *3 + 5 = 8.*
- Say the fact that goes backward along the arrow. (Signal.) *8 – 3 = 5.*
- Say the other minus fact. (Signal.) *8 – 5 = 3.*
- (Repeat until firm.)

from Lesson 51, Exercise 7

Teaching Note: The critical part of the exercise is shown with a bracket. Repeat this part, even if students respond well the first time you present the part. They need to become facile in saying the four facts in the order that starts with the first small number.

Also on Lesson 51, you introduce the number family table. It shows all families with small numbers 1 through 10.

Here's the exercise:

EXERCISE 9: NUMBER FAMILY TABLE

a. Open your workbook to the inside front cover and touch the number families. ✔
(Teacher reference:)

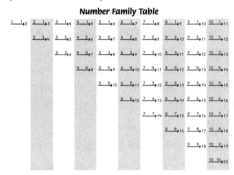

Number Family Table

(Praise children who find the right page quickly. Assist children who need help.)
This page shows the number family tables. All the families you're learning are in this table.
- Look at the top row. ✔
 ALL of the families in this row have a small number of 1.
- Touch the family with the small numbers of 1 and 1. ✔
- Touch the family with the small numbers of 2 and 1. ✔
- Touch the family with the small numbers of 3 and 1. ✔
- What are the small numbers in the next family? (Signal.) *4 and 1.*
- What are the small numbers in the next family? (Signal.) *5 and 1.*
- Touch the last family in the row. ✔
- What are the small numbers in that family? (Signal.) *10 and 1.*
b. Go back to the beginning of the next row. ✔
- The small numbers in the family are 2 and 2. All of the families in this row have a small number of 2.
- Touch the family with the small numbers of 3 and 2. ✔
- What are the small numbers in the next family? (Signal.) *4 and 2.*
- What are the small numbers in the next family? (Signal.) *5 and 2.*
- Touch the last family in the row. ✔
- What are the small numbers in that family? (Signal.) *10 and 2.*
c. All of the families below that family have a small number of 10.
 The next family down has small number of 10 and 3.
- What are the small numbers in the next family down? (Signal.) *10 and 4.*
- Touch the number family that is all the way down at the bottom. ✔
- What are the small numbers in that family? (Signal.) *10 and 10.*

Lesson 51, Exercise 9

Teaching Note: Observe students closely. Make sure they touch under each family you name. If students are slow in responding, repeat the tasks for the top row and the next row.

A cycle for teaching pairs of new number families begins on Lesson 54. The cycle has four exercise types:

1. The families are introduced. Students identify the small numbers and big numbers of the families and say all facts for each family.

2. Students identify the missing numbers in families.

3. Students identify the answers to sets of problems (all minus problems or all plus problems).

4. Students identify answers to mixed sets of minus and plus problems.

Note that the cycle presents more than one fact exercise on every lesson.

Below is a schedule for the families that are introduced on Lessons 54 through the end of the program.

Familes	Facts	Introduced on Lesson
$6 \xrightarrow{3} 9$ $6 \xrightarrow{4} 10$	$6 + 3 = 9,\ 3 + 6 = 9,\ 9 - 3 = 6,$ $9 - 6 = 3,\ 6 + 4 = 10,\ 4 + 6 = 10,$ $10 - 4 = 6,\ 10 - 6 = 4$	54
$6 \xrightarrow{5} 11$ $6 \xrightarrow{6} 12$	$6 + 5 = 11,\ 5 + 6 = 11,\ 11 - 5 = 6,$ $11 - 6 = 5,\ 6 + 6 = 12,\ 12 - 6 = 6$	64
$4 \xrightarrow{3} 7$ $4 \xrightarrow{4} 8$	$4 + 3 = 7,\ 3 + 4 = 7,\ 7 - 3 = 4,$ $7 - 4 = 3,\ 4 + 4 = 8,\ 8 - 4 = 4$	79
$3 \xrightarrow{3} 6$ $5 \xrightarrow{5} 10$	$3 + 3 = 6,\ 6 - 3 = 3,$ $5 + 5 = 10,\ 10 - 5 = 5$	89
$7 \xrightarrow{7} 14$ $8 \xrightarrow{8} 16$	$7 + 7 = 14,\ 14 - 7 = 7,$ $8 + 8 = 16,\ 16 - 8 = 8$	99
$7 \xrightarrow{3} 10$ $8 \xrightarrow{3} 11$	$7 + 3 = 10,\ 3 + 7 = 10,\ 10 - 3 = 7,$ $10 - 7 = 3,\ 8 + 3 = 11,\ 3 + 8 = 11,$ $11 - 3 = 8,\ 11 - 8 = 3$	109

Following are examples of fact-cycle exercises for the families:

$$6 \xrightarrow{\hspace{2cm} 3} 9$$

$$6 \xrightarrow{\hspace{2cm} 4} 10$$

EXERCISE 3: NUMBER FAMILIES
SMALL NUMBER OF 6 `REMEDY`

a. (Display:) [54:3A]

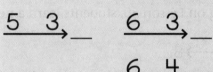

These families have a small number of 5 or a small number of 6. The big number is missing in each family. Some of these families are new.

- (Point to $6 \xrightarrow{2}$.) What are the small numbers? (Touch.) *6 (and) 2.*
- What's the big number? (Signal.) *8.*

b. (Point to $6 \xrightarrow{3}$.) What are the small numbers? (Touch.) *6 (and) 3.*
- What's the big number? (Signal.) *9.* (Touch.) *6 (and) 4.*
- What's the big number? (Signal.) *10.*

d. (Point to $6 \xrightarrow{2}$.) Say the fact that starts with the first small number. Get ready. (Signal.) *6 + 2 = 8.*
- Say the fact that starts with the other small number. Get ready. (Signal.) *2 + 6 = 8.*

e. (Point to $6 \xrightarrow{3}$.) What are the small numbers in this family? (Signal.) *6 (and) 3.*
- What's the big number? (Signal.) *9.*
- Say the fact that starts with the first small number. Get ready. (Signal.) *6 + 3 = 9.*
- Say the fact that starts with the other small number. Get ready. (Signal.) *3 + 6 = 9.*

f. (Point to $6 \xrightarrow{4}$.) What are the small numbers in this family? (Signal.) *6 (and) 4.*
- What's the big number? (Signal.) *10.*
- Say the fact that starts with the first small number. Get ready. (Signal.) *6 + 4 = 10.*
- Say the fact that starts with the other small number. Get ready. (Signal.) *4 + 6 = 10.*
(Repeat step f until firm.)

g. (Point to $5 \xrightarrow{3}$.) Here's a number family you know.
- What are the small numbers? (Touch.) *5 (and) 3.*
- What's the big number? (Signal.) *8.*
- Say the fact that starts with the first small number. Get ready. (Signal.) *5 + 3 = 8.*
- Say the fact that starts with the other small number. Get ready. (Signal.) *3 + 5 = 8.*

h. (Point to $6 \xrightarrow{3}$.) Say the fact that starts with the first small number. Get ready. (Signal.) *6 + 3 = 9.*
- Say the fact that starts with the other small number. Get ready. (Signal.) *3 + 6 = 9.*
(Repeat step h until firm.)

i. This time, you'll tell me about some minus facts.
- (Point to $5 \xrightarrow{3}$.) Say the fact that goes backward along the arrow. Get ready. (Signal.) *8 - 3 = 5.*

- Say the other minus fact. (Signal.) *8 – 5 = 3.*
j. (Point to 6─3.) Say the fact that goes backward along the arrow. Get ready. (Signal.) *9 – 3 = 6.*
- Say the other minus fact. (Signal.) *9 – 6 = 3.*
k. (Point to 6─4.) Say the fact that goes backward along the arrow. Get ready. (Signal.) *10 – 4 = 6.*
- Say the other minus fact. (Signal.) *10 – 6 = 4.*

Lesson 54, Exercise 3

Here's part of the exercise in which students identify missing numbers in number families.

Note that the example set includes the new families and families that students have learned earlier (for instance ═6→16).

g. (Display:) [56:1B]

Each family has a number missing. You'll say the problem and the missing number for each family.
- (Point to ═6→7.) Is a small number or the big number missing from this family? (Signal.) *A small number.*
- Say the problem for the missing number. Get ready. (Signal.) *7 minus 6.*
- What's the answer? (Signal.) *1.*
h. (Point to 6─4→__.) Is a small number or the big number missing from this family? (Signal.) *The big number.*
- Say the problem for the missing number. Get ready. (Signal.) *6 plus 4.*
- What's the answer? (Signal.) *10.*
i. (Repeat the following tasks for remaining families:)

(Point to __.)	Is a small number or the big number missing from this family?	Say the problem for the missing number.	What's the answer?
═6→16	A small number.	16 – 6	10
═5→8	A small number.	8 – 5	3
6─→9	A small number.	9 – 6	3
7─2→__	The big number.	7 + 2	9
6─→10	A small number.	10 – 6	4
5─3→__	The big number.	5 + 3	8

from Lesson 56, Exercise 1

For the next lessons in the cycle, students work sets of similar problems.

Here's one of the fact-cycle exercises from Lesson 60 that presents all minus problems:

EXERCISE 9: FACTS
SUBTRACTION

a. Find part 4 on worksheet 60. ✔
(Teacher reference:)

a. 10–9	d. 9–6	g. 10–8
b. 7–5	e. 8–7	h. 4–2
c. 10–4	f. 8–3	

All of these are minus problems, so they show the big number and a small number of a family. You're going to touch each problem as I read it. Then you'll tell me the answer.
- Problem A is 10 minus 9. What's the answer? (Signal.) *1.*
b. Problem B is 7 minus 5. What's the answer? (Signal.) *2.*
c. (Repeat the following tasks for problems C through H:)

Problem __ is __.		What's the answer?
C	10 – 4	6
D	9 – 6	3
E	8 – 7	1
F	8 – 3	5
G	10 – 8	2
H	4 – 2	2

d. Complete the equations in part 4. Put your pencil down when you're finished.
(Observe children and give feedback.)
(Answer key:)

a. 10–9=1	d. 9–6=3	g. 10–8=2
b. 7–5=2	e. 8–7=1	h. 4–2=2
c. 10–4=6	f. 8–3=5	

e. Check your work. You'll touch and read each fact.
- Fact A. (Signal.) *10 – 9 = 1.*
- (Repeat for:) B, *7 – 5 = 2;* C, *10 – 4 = 6;* D, *9 – 6 = 3;* E, *8 – 7 = 1;* F, *8 – 3 = 5;* G, *10 – 8 = 2;* H, *4 – 2 = 2.*

Lesson 60, Exercise 9

Note that students first say answers to all the problems. Then they write the answers. Do not present the written work until students are firm on the oral work. Note problems they miss and repeat those after going through the others.

On the same lesson, students complete a set of plus facts:

```
Part 1
a. 6+4      f. 5+10
b. 7+2      g. 2+9
c. 3+6      h. 4+6
d. 10+8     i. 5+2
e. 2+3      j. 3+5
```

Workbook Lesson 60, Part 1

On Lesson 59, students complete a mixed set of problems that plus and minus.

Here's the exercise:

```
Part 1
a. 5+3      f. 3+6
b. 8−2      g. 2+4
c. 11−9     h. 9−3
d. 6+4      i. 9+10
e. 13−3     j. 10−6
```

Workbook Lesson 59, Part 1

Four of the ten problems are from the new families.

Teaching Note: For this exercise, you direct students to read the problem; then you ask:

• Is the answer the big number or a small number?

Then students say the answer to the problem.

FACT RELATIONSHIPS

On Lesson 106, students use a strategy for figuring out unfamiliar addition and subtraction facts. This strategy has been demonstrated frequently in the introductions of families that share the same small number. If the small number in a new family is one more or one less than the corresponding number in a familiar family, students are able to figure out the big number in the family. In the same way, students are able to figure out the answer to facts that have an answer that is one more or one less than a familiar fact.

Here's part of the exercise from Lesson 106 that applies this strategy to new facts: 7 plus 3, 8 plus 3, and 9 plus 3:

EXERCISE 4: FACT RELATIONSHIPS
PLUS 3   REMEDY

a. (Display:)   [106:4A]

1 + 3	4 + 3	7 + 3
2 + 3	5 + 3	8 + 3
3 + 3	6 + 3	9 + 3

These are problems that plus 3. You know the answer to some of the problems. My turn to read the plus-3 problems and say the answers. Say the answers with me if you can keep up.
1 plus 3 equals **4**.
2 plus 3 equals **5**.
• (Repeat for 3 + 3 = **6**, 4 + 3 = **7**, 5 + 3 = **8**, 6 + 3 = **9**, 7 + 3 = **10**, 8 + 3 = **11**, 9 + 3 = **12**.)

b. Now you'll read each problem and say the answer.
• (Point to **1 + 3**.) Read the problem. (Touch.) *1 plus 3.*
• What's the answer? (Signal.) *4.*

c. (Point to **2 + 3**.) Read the problem. (Touch.) *2 plus 3.*
• What's the answer? (Signal.) *5.*

d. (Repeat the following tasks for remaining problems:)

(Point to __.) Read the problem.	What's the answer?
3 + 3	6
4 + 3	7
5 + 3	8
6 + 3	9
7 + 3	10
8 + 3	11
9 + 3	12

from Lesson 106, Exercise 4

Teaching Note: In step A, present the sequence at a reasonably fast, rhythmical pace. Maintain the same pace and rhythm when students respond with you. Make it seem like fun. Repeat step A at least one more time.

The rationale for presenting in this manner is that the students will learn the material much better if it is in a predictable pattern. If students require think time to respond, the pattern is lost.

Later, students do variations of patterns. They do a pattern for minus-3 problems (4 – 3 through 13 – 3). They do a pattern for 3-plus problems (3 + 1 through 3 + 10).

Here's the exercise from Lesson 117. Students do all three patterns—first 3 plus, then minus 3, then plus 3:

EXERCISE 1: FACT RELATIONSHIPS
ADDITION AND SUBTRACTION   `REMEDY`

  a. You'll say facts for families that have a small number of 3. First you'll start with 3 plus 1 and say the 3-plus facts to 3 plus 10.
  - Say the fact for 3 plus 1. Get ready. (Signal.) *3 + 1 = 4.*
  - Next fact. Get ready. (Signal.) *3 + 2 = 5.*
  - (Repeat for:) *3 + 3 = 6, 3 + 4 = 7, 3 + 5 = 8, 3 + 6 = 9, 3 + 7 = 10, 3 + 8 = 11, 3 + 9 = 12, 3 + 10 = 13.*
  - (Repeat until firm.)
  b. Now you'll start with 4 minus 3 and say the minus-3 facts to 13 minus 3.
  - Say the fact for 4 minus 3. (Signal.) *4 – 3 = 1.*
  - Say the minus-3 fact that starts with 5. (Signal.) *5 – 3 = 2.*
  - Next fact. (Signal.) *6 – 3 = 3.*
  - (Repeat for:) *7 – 3 = 4, 8 – 3 = 5, 9 – 3 = 6, 10 – 3 = 7, 11 – 3 = 8, 12 – 3 = 9, 13 – 3 = 10.*
  - (Repeat until firm.)
  c. Now you'll start with 1 plus 3 and say the plus-3 facts to 10 plus 3.
  - Say the fact for 1 plus 3. Get ready. (Signal.) *1 + 3 = 4.*
  - Next fact. (Signal.) *2 + 3 = 5.*
  - (Repeat for:) *3 + 3 = 6, 4 + 3 = 7, 5 + 3 = 8, 6 + 3 = 9, 7 + 3 = 10, 8 + 3 = 11, 9 + 3 = 12, 10 + 3 = 13.*
  - (Repeat until firm.)
  d. Now I'll mix some of the problems up and you'll tell me the answer.
  - Listen: 3 plus 8. What's 3 plus 8? (Signal.) *11.*
  - Listen: 10 minus 3. What's 10 minus 3? (Signal.) *7.*
  - Listen: 4 plus 3. What's 4 plus 3? (Signal.) *7.*
  - Listen: 11 minus 3. What's 11 minus 3? (Signal.) *8.*

▬▬▬▬▬ **INDIVIDUAL TURNS** ▬▬▬▬▬
(Call on individual students to perform one of the following tasks.)

  - What's 11 minus 3? (Call on a student.) *8.*
  - What's 4 plus 3? (Call on a student.) *7.*
  - What's 10 minus 3? (Call on a student.) *7.*
  - What's 3 plus 8? (Call on a student.) *11.*

Lesson 117, Exercise 1

From Lessons 121 through the end of the level, students work on similar patterns of related facts. They work on patterns for 4 + 5 through 4 + 10 and 5 + 5 through 5 + 10.

USING NUMBER FAMILIES TO SOLVE PROBLEMS

This track focuses on problems that require addition or subtraction calculations. Some problems simply require students to work addition or subtraction problems based on information provided by number families. Some problems are in the form of a word problem. Some problems require students to make number families for addition or subtraction problems.

The first exercises show students that the principles they have applied to number families for facts apply to families that have larger numbers. Each family has a big number and two small numbers. If the big number is missing, you add the small numbers. If a small number is missing, you start with the big number and subtract the small number that is shown. The most basic property of families is that you can formulate two addition facts and two subtraction facts.

This property is taught on Lesson 91. Here's the introduction:

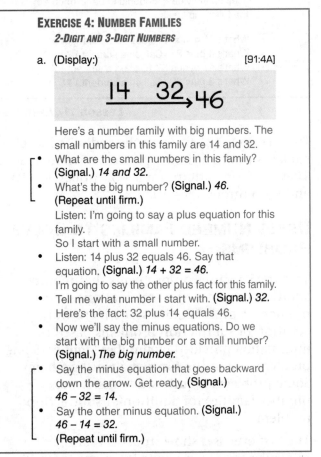

> **EXERCISE 4: NUMBER FAMILIES**
> *2-DIGIT AND 3-DIGIT NUMBERS*
>
> a. (Display:) [91:4A]
>
> $$\underline{14 \quad 32}_{\,\searrow}46$$
>
> Here's a number family with big numbers. The small numbers in this family are 14 and 32.
> • What are the small numbers in this family? (Signal.) *14 and 32.*
> • What's the big number? (Signal.) *46.* (Repeat until firm.)
> Listen: I'm going to say a plus equation for this family.
> So I start with a small number.
> • Listen: 14 plus 32 equals 46. Say that equation. (Signal.) *14 + 32 = 46.*
> I'm going to say the other plus fact for this family.
> • Tell me what number I start with. (Signal.) *32.*
> Here's the fact: 32 plus 14 equals 46.
> • Now we'll say the minus equations. Do we start with the big number or a small number? (Signal.) *The big number.*
> • Say the minus equation that goes backward down the arrow. Get ready. (Signal.) *46 – 32 = 14.*
> • Say the other minus equation. (Signal.) *46 – 14 = 32.* (Repeat until firm.)

from Lesson 91, Exercise 4

After students have worked similar exercises for several lessons, they make number families for equations.

Here's part of the exercise from Lesson 101:

> c. We're going to write number families for these equations.
> • (Point to **63**.) Read the equation. (Touch.) *63 + 12 = 75.*
> • Does the equation add or subtract? (Signal.) *Add.*
> • So what's the big number? (Signal.) *75.*
> • 75 is the big number. What are the small numbers? (Signal.) *63 and 12.*
> I write the big number first and then the small numbers.
> (Add to show:) [101:5B]
>
> $$\begin{array}{r} 63 \\ +\;12 \\ \hline 75 \end{array}$$
>
> $$\underline{63 \quad 12}_{\,\searrow}75$$
>
> Here's the family.

from Lesson 101, Exercise 5

Starting on Lesson 104, students work a variation of this problem type. The answers to the problems are not shown. So students first work the problem, then they make the number family with three numbers.

Here are the problems from Lesson 104:

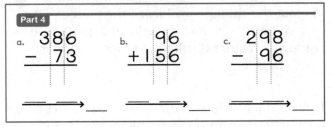

> **Part 4**
>
> a. $\begin{array}{r} 386 \\ -\;73 \\ \hline \end{array}$ b. $\begin{array}{r} 96 \\ +156 \\ \hline \end{array}$ c. $\begin{array}{r} 298 \\ -\;96 \\ \hline \end{array}$

Workbook Lesson 104, Part 4

Starting on Lesson 106, students work difficult problems that require them to pair written problems with one of two number families.

Here are the problems from Lesson 110:

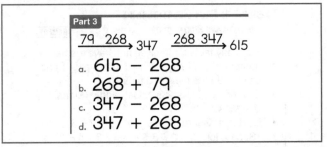

> **Part 3**
>
> $\underline{79 \quad 268}_{\,\searrow}347$ $\underline{268 \quad 347}_{\,\searrow}615$
>
> a. $615 - 268$
> b. $268 + 79$
> c. $347 - 268$
> d. $347 + 268$

Workbook Lesson 110, Part 3

Students do not have to solve the problems shown. Rather, they identify the correct number family and then copy the missing number in the problem.

Here are the families and problems from Lesson 110. Note that two numbers—268 and 347—occur in both families. Therefore, students must determine whether the number shown in a problem is the big number or a small number. For example: 347 – 268.

The problem subtracts. So the big number is 347. Students find the family with the big number of 347. The missing number is 79.

Starting on Lesson 113 students work with equations that have a missing number in any of three positions. Students identify whether the missing number is the big number or a small number. They then pair the numbers in the problem with number families and complete the equation.

Here are the problems from Lesson 113:

Part 2

a. $243 - \underline{\quad} = 89$ c. $\underline{\quad} - 65 = 178$
b. $\underline{\quad} + 89 = 154$ d. $243 - 178 = \underline{\quad}$

Workbook Lesson 113, Part 2

Students identify that the missing number in A is a small number. The big number in the problem is 243. So they refer to the families shown below to find a number family that has a big number of 243 and a small number of 89.

e. (Display:)   [113:5A]

1. $\underset{178 \quad 65}{\xrightarrow{\hspace{2cm}}} 243$

2. $\underset{89 \quad 154}{\xrightarrow{\hspace{2cm}}} 243$

3. $\underset{65 \quad 89}{\xrightarrow{\hspace{2cm}}} 154$

from Lesson 113, Exercise 5

The missing number in the problem is 154.

Starting on Lesson 119 students apply what they have learned to solve word problems with unknowns in all positions. They learn that if a problem tells about *some*, the number is not known. So you write a space ___ for that number, and you have to figure out that number. For example:

> A carpenter started out with some boards. Then the carpenter bought 6 more boards. The carpenter ended up with 28 boards. How many boards did he start out with?

The first sentence does not give a number so students write a space.

Here's what students write for the word problem: ___ + 6 = 28.

They recognize that 28 is the big number and 6 is a small number, so they write these numbers in the family.

$$\underset{\underline{\quad} \quad 6}{\xrightarrow{\hspace{2cm}}} 28$$

Then they work the problem 28 – 6.

Here's part of an exercise in which students solve a problem like the one above:

a. Find part 4 on your worksheet. ✔
    (Teacher reference:)

a. _____ b. _____

I'll tell you word problems. You'll write the symbols for each problem. Then you'll make the number family for the problem and figure out the answer.
Problem A: A board was 11 feet long. A carpenter cut part of the board off. The board ended up 6 feet long. How much of the board was cut off?

• Listen to the first part again: A board was 11 feet long. What number do you write for that part? (Signal.) *11.*
• Write it. ✔
• Next part: A carpenter cut part of the board off. Tell me the symbols you write for that part. (Signal.) *Minus space.*
• Write the minus sign and the space. ✔
b. Next part: The board ended up 6 feet long. Tell me the symbols for that part. (Signal.) *Equals 6.*
• Write equals 6. ✔
    (Display:)   [122:7A]

a. $11 - \underline{\quad} = 6$

Here's what you should have for problem A.
• Read the problem you wrote for A. Get ready. (Signal.) *11 minus how many equals 6.*
c. Look at the problem you wrote for A and get ready to tell me about the numbers.
• Is the missing number the big number or a small number? (Signal.) *A small number.*
• What is 11? (Signal.) *The big number.*
• What is 6? (Signal.) *A small number.*
• Make the family for A.
    (Observe children and give feedback.)
    (Add to show:)   [122:7B]

a. $11 - \underline{\quad} = 6$

$$\underset{6}{\xrightarrow{\hspace{1.5cm}}} 11$$

Here's what you should have.
• Say the problem for the missing number in family A. (Signal.) *11 minus 6.*
• What's the answer? (Signal.) *5.*
    5 tells how many feet of the board was cut off.
• Listen to problem A again: A board was 11 feet long. A carpenter cut part of the board off. The board ended up 6 feet long. How much of the board was cut off?
• Everybody, what's the answer? (Signal.) *5 (feet).*

from Lesson 122, Exercise 7

Students who complete the track have a good understanding of addition and subtraction problems that have missing numbers in any of the three positions.

Column Addition and Subtraction

Students who have gone through *CMC Level A* have learned effective procedures for solving two-digit problems, including those that traditionally require carrying. They solve these problems by applying a place-value analysis of the numbers in the problem.

For this problem:

$$\begin{array}{r} 27 \\ +43 \\ \hline \end{array}$$

students make Ts and lines for 43. Each T represents a 10; each line represents a 1.

So the Ts and lines for 43 are T T T T I I I.

$$\begin{array}{r} 27 \\ +43 \\ \hline \end{array} \text{ T T T T I I I}$$

Now students start with the top number and count on for the Ts and the lines.

They say 27 (pause) 37, 47, 57, 67, 68, 69, 70.

Students do a parallel solution for subtraction problems. They make Ts and lines for the first number, then cross out the Ts and lines that are subtracted.

The remaining Ts and lines show the answer to the problem.

Because the students already know how to solve these problems, some will be tempted to use a solution that involves Ts and lines.

Tell them that from now on, they will not make Ts and lines to solve those problems. Rather they will learn a new way. They will first work the problem for the ones and write the answer, then work the problem for the tens and write the answer.

CMC Level B teaches students to solve the following types of addition problems:

- Basic column addition (no carrying)

$$\begin{array}{r} 3 \\ +10 \\ \hline \end{array} \qquad \begin{array}{r} 45 \\ +34 \\ \hline \end{array}$$

- Column addition with carrying

$$\begin{array}{r} 29 \\ +42 \\ \hline \end{array} \qquad \begin{array}{r} 78 \\ +44 \\ \hline \end{array}$$

- Problems with 3 addends

$$\begin{array}{r} 7 \\ 3 \\ +5 \\ \hline \end{array} \qquad \begin{array}{r} 23 \\ 5 \\ +24 \\ \hline \end{array}$$

CMC Level B also teaches students to solve basic subtraction problems (no borrowing).

$$\begin{array}{r} 46 \\ -10 \\ \hline \end{array} \qquad \begin{array}{r} 84 \\ -14 \\ \hline \end{array} \qquad \begin{array}{r} 97 \\ -35 \\ \hline \end{array}$$

Basic Column Addition and Subtraction

The first two-digit column problems are introduced on Lesson 27.

Here's part of the exercise:

EXERCISE 11: COLUMN PROBLEMS
ADDITION AND SUBTRACTION   `REMEDY`

a. (Display:)   [27:11A]

$$\begin{array}{r} 46 \\ -10 \\ \hline \end{array} \qquad \begin{array}{r} 75 \\ +21 \\ \hline \end{array} \qquad \begin{array}{r} 84 \\ -14 \\ \hline \end{array}$$

You'll read each problem.
- (Point to **46**.) Read the problem. (Touch.) *46 minus 10.*
- (Point to **75**.) Read the problem. (Touch.) *75 plus 21.*
- (Point to **84**.) Read the problem. (Touch.) *84 minus 14.*

b. (Point to **84**.) This problem is 84 minus 14.
    (Point to $-\frac{84}{14}$.) My turn to read the problem for the ones column. (Touch symbols.) *4 minus 4.*
- Your turn: Read the problem for the ones column. Get ready. (Touch symbols.) *4 minus 4.*
- (Point to $-\frac{84}{14}$.) Read the problem for the tens column. Get ready. (Touch symbols.) *8 minus 1.*
  (Repeat step b until firm.)

c. (Point to **75**.) Read the whole problem. Get ready. (Touch.) *75 plus 21.*
- (Point to $+\frac{75}{21}$.) Read the problem for the ones. Get ready. (Touch.) *5 plus 1.*
- (Point to $+\frac{75}{21}$.) Read the problem for the tens. Get ready. (Touch.) *7 plus 2.*

---

d. (Point to **46.**) Read the whole problem. Get ready. (Touch.) *46 minus 10.*

- (Point to ⁻¹₀⁶.) Read the problem for the ones. Get ready. (Touch.) *6 minus zero.*

- (Point to ⁻¹₀⁶.) Read the problem for the tens. Get ready. (Touch.) *4 minus 1.*
(Repeat tasks in steps c and d that were not firm.)

e. Find part 4 on your worksheet. ✔
(Teacher reference:)　　　　　　R Part L

```
 a. | 4 | 6      b. | 7 | 5      c. | 8 | 4
  – | 1 | 0       + | 2 | 1       – | 1 | 4
```

These are the same problems you just read.

- Touch and read problem A. Get ready. (Signal.) *46 minus 10.*
- Touch and read the problem for the ones. Get ready. (Signal.) *6 minus zero.*
- What's the answer to 6 minus zero? (Signal.) *6.*
- Write the answer in the ones column below the zero.
(Observe children and give feedback.)
- Touch and read the problem for the tens. Get ready. (Signal.) *4 minus 1.*
- What's the answer to 4 minus 1? (Signal.) *3.*
- Write the answer for the tens column below the 1.
(Observe children and give feedback.)
- Everybody, touch and read the whole equation for A. Get ready. (Signal.) *46 – 10 = 36.*
(Repeat until firm.)

f. Touch and read problem B. Get ready. (Signal.) *75 plus 21.*
- Touch and read the problem for the ones. Get ready. (Signal.) *5 plus 1.*
- What's the answer to 5 plus 1? (Signal.) *6.*
- Write the answer for the ones column below the 1.
(Observe children and give feedback.)
- Touch and read the problem for the tens. Get ready. (Signal.) *7 plus 2.*
- What's the answer to 7 plus 2? (Signal.) *9.*
- Write the answer in the tens column below the 2.
(Observe children and give feedback.)
- Everybody, touch and read the equation for B. Get ready. (Signal.) *75 + 21 = 96.*
(Repeat step f until firm.)

from Lesson 27, Exercise 11

Teaching Note: Before Lesson 27, students have worked column problems that present familiar facts.

```
  1 4        9       1 0
 –  4       + 1     + 7
```

The practice in step A assures that students are firm in reading different problems.

- First they read the problem.
- Then they read the problem for the ones.
- Then they read the problem for the tens.

Students have to listen carefully to your directions in steps C and D.

Do not present the Workbook part until students are firm on steps C and D.

The problems presented in the Workbook (steps E and F) are the same problems presented in the preceding steps.

Monitor students to confirm that they are following the directions "Touch and read _____".

Three-digit problems are introduced on Lesson 30.

Here's the exercise:

from Lesson 30, Exercise 4

The only difference between these problems and the ones students have been working is that they work the problem for the hundreds column after working the problem for the tens column.

Starting on Lesson 35 students work a new kind of problem. The numbers in the problem do not have the same number of digits. One number may have one digit and the other three digits.

Here's part of the exercise from Lesson 38:

from Lesson 38, Exercise 7

Teaching Note: The routine is for students to read the whole problem, then read the problem for the ones. If there are two numbers in the column for the tens, students read the problem. If there is only one number in a column, students indicate the number that goes in the answer.

After going through the problems verbally, students write answers to the problems and then check their work.

Three Addends

On Lesson 36, students work problems that add three numbers. Adding three numbers is practiced before students learn to carry, which requires adding three numbers in the tens or hundreds column.

The part that adds the first two numbers is underlined. Students read the underlined part and say the answer. Then they say the problem that starts with the answer.

Here's the first part of the exercise from Lesson 40:

> **EXERCISE 4: ADDITION**
> *3 ADDENDS*
>
> a. (Display:) W̅   [40:4A]
>
> $$\underline{6 + 2} + 1$$
>
> • (Point to **6**.) Read the whole problem. (Touch.) *6 plus 2 plus 1.*
> • Say the problem for the underlined part. (Touch.) *6 plus 2.*
> • What's the answer? (Signal.) *8.*
> • Say the next problem. (Signal.) *8 plus 1.*
> • (Repeat until firm.)
> • What's the answer? (Signal.) *9.*
>   (Add to show:)   [40:4B]
>
> $$\underline{6 + 2} + 1 = 9$$
>
> • Read the equation. (Touch.) *6 + 2 + 1 = 9.*

from Lesson 40, Exercise 4

Teaching Note: If students have difficulty responding to the direction, "Say the next problem," tell them to start with 8 and say the next problem.

Then repeat the bracketed part without this prompt.

In later lessons, students work these problems in their Workbooks. The first part of these problems is not underlined.

A column variation of these problems starts on Lesson 51.

Here are the problems from that lesson:

> a. $\begin{array}{r} 2 \\ 8 \\ + 4 \\ \hline \end{array}$ c. $\begin{array}{r} 3 \\ 2 \\ + 1 \\ \hline \end{array}$
>
> b. $\begin{array}{r} 1 \\ 6 \\ + 1 \\ \hline \end{array}$ d. $\begin{array}{r} 1 \\ 9 \\ + 3 \\ \hline \end{array}$

from Lesson 51, Exercise 4

On Lesson 64, students work column problems that add three 2-digit numbers.

COLUMN SUBTRACTION – ZERO AS FIRST DIGIT OF ANSWER

On Lesson 54, students learn a skill that is needed to work subtraction problems of the form:

$$\begin{array}{r} 386 \\ -321 \\ \hline \end{array}$$

Students learn the convention that zero cannot be the digit a number starts with. So the answer to the problem is 65, not 065.

On Lesson 55, students work problems of this form.

Here's the first part of the exercise:

> **EXERCISE 6: COLUMN SUBTRACTION**
> *WHEN 1ST EQUALS ZERO*   [REMEDY]
>
> a. (Display:) W̅   [55:6A]
>
> $$\begin{array}{r} 53 \\ -51 \\ \hline \end{array} \quad \begin{array}{r} 439 \\ -410 \\ \hline \end{array} \quad \begin{array}{r} 64 \\ -60 \\ \hline \end{array}$$
>
> The beginning digit of the answer for these problems is zero. But we're going to write the right digits for the answer.
> • (Point to **53**.) Read the problem. Get ready. (Touch.) *53 minus 51.*
> • Read the problem for the ones. (Touch.) *3 minus 1.*
> • What's the answer? (Signal.) *2.*
>   (Add to show:)   [55:6B]
>
> $$\begin{array}{r} 53 \\ -51 \\ \hline 2 \end{array} \quad \begin{array}{r} 439 \\ -410 \\ \hline \end{array} \quad \begin{array}{r} 64 \\ -60 \\ \hline \end{array}$$
>
> • Read the problem for the tens. Get ready. (Touch.) *5 minus 5.*
> • What's the answer? (Signal.) *Zero.*
>   (Add to show:)   [55:6C]
>
> $$\begin{array}{r} 53 \\ -51 \\ \hline 02 \end{array} \quad \begin{array}{r} 439 \\ -410 \\ \hline \end{array} \quad \begin{array}{r} 64 \\ -60 \\ \hline \end{array}$$
>
> (Point to **0**.) Is the number right or wrong? (Touch.) *Wrong.*
> (Erase to show:)   [55:6D]
>
> $$\begin{array}{r} 53 \\ -51 \\ \hline 2 \end{array} \quad \begin{array}{r} 439 \\ -410 \\ \hline \end{array} \quad \begin{array}{r} 64 \\ -60 \\ \hline \end{array}$$
>
> Now the answer is right.
> • Read the whole equation. (Touch.) *53 – 51 = 2.*

from Lesson 55, Exercise 6

Column Addition with Carrying

On Lesson 73, the work prepares students with additional skills and operations that are used when they carry.

The first skill is writing a 2-digit answer to the problem in the ones column.

For example:

$$\begin{array}{r} 2\,6 \\ +3\,4 \\ \hline \end{array}$$

Students say the problem for the ones—6 plus 4. The answer is 10. They write the tens digit in the tens column and the ones digit in the ones column.

$$\begin{array}{r} {\scriptstyle 1} \\ 2\,6 \\ +3\,4 \\ \hline 0 \end{array}$$

The next isolated skill students practice is reading the new problem in the tens column (1 + 2 + 3).

On Lesson 76, students work all the steps in problems that carry.

Here's part of the exercise from Lesson 76:

EXERCISE 2: COLUMN ADDITION
CARRYING

a. If the answer for the ones column has two digits, you write the tens digit in the tens column and the ones digit in the ones column.
- Where do you write the ones digit? (Signal.) *In the ones column.*
- Where do you write the tens digit? (Signal.) *In the tens column.*

b. (Display:) W   [76:2A]

$$\begin{array}{r} 7\,2 \\ +1\,8 \\ \hline \end{array}$$

- (Point to **72**.) Read the problem. (Touch.) *72 plus 18.*
- Say the problem for the ones. (Signal.) *2 plus 8.*
- What's the answer? (Signal.) *10.*
- What's the tens digit of 10? (Signal.) *1.*
- What's the ones digit of 10? (Signal.) *Zero.*
- Where do I write the tens digit? (Signal.) *In the tens column.*
- Where do I write the ones digit? (Signal.) *In the ones column.*
  (Add to show:)   [76:2B]

$$\begin{array}{r} {\scriptstyle 1} \\ 7\,2 \\ +1\,8 \\ \hline 0 \end{array}$$

c. Read the new problem in the tens column. (Touch.) *1 plus 7 plus 1.*
- What's 1 plus 7? (Signal.) *8.*
- What's 8 plus 1? (Signal.) *9.*
  That's the answer for the tens column.
  (Add to show:)   [76:2C]

$$\begin{array}{r} {\scriptstyle 1} \\ 7\,2 \\ +1\,8 \\ \hline 9\,0 \end{array}$$

- Look at the answer. What's 72 plus 18? (Signal.) *90.*
- Read the problem and the answer. (Signal.) *72 + 18 = 90.*

from Lesson 76, Exercise 2

On Lesson 77, students work carrying problems in their Workbooks.

Here's the exercise from Lesson 79:

EXERCISE 4: COLUMN ADDITION
CARRYING

a. (Distribute unopened workbooks to children.)
- Open your workbook to Lesson 79 and find part 1.
(Observe children and give feedback.)
(Teacher reference:)

$$\begin{array}{r} a.\quad 44 \\ +\ 36 \\ \hline \end{array}$$

$$\begin{array}{r} b.\quad 16 \\ +\ 76 \\ \hline \end{array}$$

- Touch and read problem A. (Signal.) *44 plus 36.*
- Read the problem in the ones column. (Signal.) *4 plus 6.*
- What's the answer? (Signal.) *10.*
- What's the tens digit of 10? (Signal.) *1.*
- What's the ones digit of 10? (Signal.) *Zero.*
- Write the digits for 10 where they should go. (Observe children and give feedback.)
b. Touch and read the new problem for the tens column. (Signal.) *1 plus 4 plus 3.*
- What's 1 plus 4? (Signal.) *5.*
- What's 5 plus 3? (Signal.) *8.*
- Write 8 where it should go. (Observe children and give feedback.)
- Start with 44 and read the problem and the answer. (Signal.) *44 + 36 = 80.*
c. Touch and read problem B. (Signal.) *16 plus 76.*
- Read the problem in the ones column. (Signal.) *6 plus 6.*
- What's the answer? (Signal.) *12.*
- What's the tens digit of 12? (Signal.) *1.*
- What's the ones digit of 12? (Signal.) *2.*
- Write the digits for 12 where they should go. (Observe children and give feedback.)
d. Touch and read the problem for the tens column. (Signal.) *1 plus 1 plus 7.*
- What's 1 plus 1? (Signal.) *2.*
- What's 2 plus 7? (Signal.) *9.*
- Write 9 where it should go. (Observe children and give feedback.)
- Start with 16 and read the problem and the answer. (Signal.) *16 + 76 = 92.*

---

from Lesson 79, Exercise 4

Teaching Note: The steps that students take to work the problem have been practiced one at a time and in combinations. With this practice, students should have no problem combining the steps to work the problem presented in this exercise.

Starting on Lesson 82, students work problem sets in which some of the problems require carrying, others don't.

Here's the first part of the exercise.

EXERCISE 7: COLUMN ADDITION
CARRYING DISCRIMINATION   REMEDY

a. Find part 3 on worksheet 82. ✔
(Teacher reference:)   R Part J

$$\begin{array}{ccc} a.\ 11 & b.\ 37 & c.\ 38 \\ +59 & +42 & +42 \\ \hline \end{array}$$

You have to carry to work some of these problems.
- Problem A is 11 plus 59. Say the problem for the ones. Get ready. (Signal.) *1 plus 9.*
- What's the answer? (Signal.) *10.*
- Do you write anything in the tens column? (Signal.) *Yes.*
b. Problem B is 37 plus 42. Say the problem for the ones. Get ready. (Signal.) *7 plus 2.*
- What's the answer? (Signal.) *9.*
- Do you write anything in the tens column? (Signal.) *No.*
c. Problem C is 38 plus 42. Say the problem for the ones. Get ready. (Signal.) *8 plus 2.*
- What's the answer? (Signal.) *10.*
- Do you write anything in the tens column? (Signal.) *Yes.*
d. Work all the problems in part 3. (Observe children and give feedback.)

---

from Lesson 82, Exercise 7

Teaching Note: The key question for each problem comes after students say the answer to the problem for the ones. The question is: "Do you write anything in the tens column?"

If students make a mistake ask, "Is the answer a 2-digit number? So do you write anything in the tens column?"

Problems that carry digits to the hundreds column first appear on Lesson 86.

Here's the problem set from Lesson 86:

EXERCISE 3: COLUMN ADDITION
CARRYING WITH 3-DIGIT NUMBERS   `REMEDY`

a. (Display:) W   [86:3A]

$$\begin{array}{r} 2\,6\,7 \\ +\,3\,6\,1 \\ \hline \end{array}$$

- Everybody, read the problem. Get ready.
  (Signal.) *267 plus 361.*
- Read the problem for the ones. Get ready.
  (Signal.) *7 plus 1.*
- What's the answer? (Signal.) *8.*
  (Add to show:)   [86:3B]

$$\begin{array}{r} 2\,6\,7 \\ +\,3\,6\,1 \\ \hline 8 \end{array}$$

- Read the problem for the tens. Get ready.
  (Signal.) *6 plus 6.*
- What's the answer? (Signal.) *12.*
- What are the digits of 12? (Signal.) *1 and 2.*
  So I write 1 in the hundreds column and 2 in
  the tens column.
  Watch.
  (Add to show:)   [86:3C]

$$\begin{array}{r} {}^{1}\,2\,6\,7 \\ +\,3\,6\,1 \\ \hline 2\,8 \end{array}$$

- Read the problem for the hundreds. Get ready.
  (Signal.) *1 plus 2 plus 3.*
- Tell me the answer. Get ready. (Signal.) *6.*
  (Add to show:)   [86:3D]

$$\begin{array}{r} {}^{1}\,2\,6\,7 \\ +\,3\,6\,1 \\ \hline 6\,2\,8 \end{array}$$

- What's 267 plus 361? (Signal.) *628.*

---

b. (Display:)   [86:3E]

$$\begin{array}{r} 1\,6\,6 \\ +\,3\,2\,5 \\ \hline \end{array}$$

- Everybody, read the problem. Get ready.
  (Signal.) *166 plus 325.*
- Read the problem for the ones. Get ready.
  (Signal.) *6 plus 5.*
- What's the answer? (Signal.) *11.*
- What do I write in the tens column? (Signal.) *1.*
- What do I write in the ones column? (Signal.) *1.*
  (Add to show:)   [86:3F]

$$\begin{array}{r} {}^{1}\,1\,6\,6 \\ +\,3\,2\,5 \\ \hline 1 \end{array}$$

- Read the problem for the tens. Get ready.
  (Signal.) *1 plus 6 plus 2.*
- Tell me the answer. Get ready. (Signal.) *9.*
  (Add to show:)   [86:3G]

$$\begin{array}{r} {}^{1}\,1\,6\,6 \\ +\,3\,2\,5 \\ \hline 9\,1 \end{array}$$

- Read the problem for the hundreds. Get ready.
  (Signal.) *1 plus 3.*
- What's the answer? (Signal.) *4.*
  (Add to show:)   [86:3H]

$$\begin{array}{r} {}^{1}\,1\,6\,6 \\ +\,3\,2\,5 \\ \hline 4\,9\,1 \end{array}$$

- What's 166 plus 325? (Signal.) *491.*

Lesson 86, Exercise 3

These problems are of the same form. If the answer in the ones column or the tens column is a 2-digit number, you write the digits in the appropriate columns.

On Lesson 90, students work problems that require carrying in both the tens and hundreds column.

Problems that require carrying in two columns simply repeat the carrying operation for the hundreds digits.

Here's the first part of the exercise from Lesson 90:

EXERCISE 2: COLUMN ADDITION
CARRYING (MULTIPLE DIGITS)　　　　REMEDY

a. Here's a new kind of problem. You have to carry a digit from the ones column and carry a digit from the tens column.
(Display:) W　　　　　　　　　[90:2A]

$$\begin{array}{r} 2\ 9\ 6 \\ +\ 5\ 4\ 6 \\ \hline \end{array}$$

- Read the problem. Get ready. (Signal.) *296 plus 546.*
- Read the problem for the ones. (Signal.) *6 plus 6.*
- What's the answer? (Signal.) *12.*
  That answer has two digits, 1 and 2. We write 2 in the ones column.
  (Add to show:)　　　　　　　[90:2B]

$$\begin{array}{r} 2\ 9\ 6 \\ +\ 5\ 4\ 6 \\ \hline 2 \end{array}$$

- What do we write in the tens column? (Signal.) *1.*
  (Add to show:)　　　　　　　[90:2C]

$$\begin{array}{r} {\scriptstyle 1} \\ 2\ 9\ 6 \\ +\ 5\ 4\ 6 \\ \hline 2 \end{array}$$

- Read the problem for the tens column. Get ready. (Signal.) *1 plus 9 plus 4.*
- What's 1 plus 9? (Signal.) *10.*
- What's 10 plus 4? (Signal.) *14.*
  That answer has two digits, 1 and 4. We write 4 in the tens column.
  (Add to show:)　　　　　　　[90:2D]

$$\begin{array}{r} {\scriptstyle 1} \\ 2\ 9\ 6 \\ +\ 5\ 4\ 6 \\ \hline 4\ 2 \end{array}$$

- What do we write in the hundreds column? (Signal.) *1.*
  (Add to show:)　　　　　　　[90:2E]

$$\begin{array}{r} {\scriptstyle 1}\ {\scriptstyle 1} \\ 2\ 9\ 6 \\ +\ 5\ 4\ 6 \\ \hline 4\ 2 \end{array}$$

- Read the problem for the hundreds. Get ready. (Signal.) *1 plus 2 plus 5.*
- What's the answer? (Signal.) *8.*
  (Add to show:)　　　　　　　[90:2F]

$$\begin{array}{r} {\scriptstyle 1}\ {\scriptstyle 1} \\ 2\ 9\ 6 \\ +\ 5\ 4\ 6 \\ \hline 8\ 4\ 2 \end{array}$$

- Read the problem and the answer. Get ready. (Signal.) *296 + 546 = 842.*

from Lesson 90, Exercise 2

Teaching Note: You present minimum directions because students are familiar with the details of each step.

Different Digit Problems

On following lessons, students work sets of addition and subtraction problems of different types. Some addition problems require carrying. Some have three addends. Some require carrying more than once.

Word Problems

ACTION AND CLASSIFICATION WORD PROBLEMS

In *CMC Level A,* students learned to solve basic word problems that told about action or told about classes.

Here's an example of an action problem:

> Jamie started out with 11 dollars. He earned 5 more dollars. How many dollars did he end up with?

Here's an example of a classification problem:

> There were 4 red ducks on the lake. There were 3 black ducks on the lake. How many ducks were on the lake altogether?

Students worked these by writing a row problem: 11 + 5 =, then solved the problem by making Ts and lines.

In *CMC Level B,* students use different computation strategies. The initial problems involve facts that students have learned. Students write the problems in a row (without making Ts and lines) and write the answer. They work problems by adding or subtracting.

Here's part of the exercise from Lesson 27:

---

b. Listen to problem B: 30 leaves were on a plant. Then 6 more leaves grew on the plant. How many leaves ended up on the plant?

- Listen to the first part again: 30 leaves were on a plant. What symbol will you write for that part? (Signal.) *30.*
- Write the symbol for the first part of problem B. ✔
- Listen to the next part: Then 6 more leaves grew on the plant. What symbols will you write for that part? (Signal.) *Plus 6.*
- Write the symbols and complete the equation. Raise your hand when you know how many leaves ended up on the plant. **(Observe children and give feedback.)**
- Everybody, touch and read the equation for problem B. (Signal.) *30 + 6 = 36.*
- How many leaves ended up on the plant? (Signal.) *36.*

---

c. Listen to problem C: Debbie made 15 sandwiches. People ate 15 of those sandwiches. How many sandwiches were left?

- Listen to the first part again: Debbie made 15 sandwiches. What symbol will you write for that part? (Signal.) *15.*
- Write the symbol for the first part of problem C. ✔
- Listen to the next part: People ate 15 of those sandwiches. What symbols will you write for that part? (Signal.) *Minus 15.*
- Write the symbols and complete the equation. Raise your hand when you know how many sandwiches were left. **(Observe children and give feedback.)**
- Everybody, touch and read the equation for problem C. (Signal.) *15 – 15 = 0.*
- How many sandwiches were left? (Signal.) *Zero.*

---

from Lesson 27, Exercise 10

Teaching Note: Throughout this level, problems are read to students. The program does not assume that all students have skills they need to read these problems until second grade.

Make sure that students write complete equations, with the sign for adding or subtracting and the equal sign. Note that some students tend to leave off the equals sign. Correct this mistake. Also make sure that when students touch and read the equation they say all the symbols.

Starting on Lesson 44, students write equations in columns, not rows. By this lesson, students have practiced writing row problems like 35 – 22 = and 24 + 12 = as column problems:

$$
\begin{array}{r} 35 \\ -22 \\ \hline \end{array} \qquad \begin{array}{r} 24 \\ +12 \\ \hline \end{array}
$$

Note that the teacher refers to the line as the equals bar.

Here's part of the exercise from Lesson 44:

EXERCISE 7: WORD PROBLEMS
WRITING IN COLUMNS   REMEDY

a. Find part 2 on worksheet 44. ✔
(Teacher reference:)   R Part C

You're going to write the symbols for word
problems in columns and work them. The
equals bars are already shown.
- Touch where you'll write the symbols for
  problem A. ✔
  Listen to problem A: A man started with 62
  nails. The man found 14 more nails. How
  many nails did the man end up with?
- Listen again: A man started with 62 nails.
  How many nails did the man start out with?
  (Signal.) *62.*
- So what number do you start with? (Signal.) *62.*
- Does 62 start with hundreds, tens, or ones?
  (Signal.) *Tens.*
- Write 62.
  (Observe children and give feedback.)
b. Listen to the next part: The man found 14
  more nails. How many nails did the man find?
  (Signal.) *14.*
- Tell me the sign and the number you write
  for: The man found 14 more nails. Get ready.
  (Signal.) *Plus 14.*
- Write the plus sign. ✔
- Does 14 start with hundreds, tens, or ones?
  (Signal.) *Tens.*
- Write 14.
  (Observe children and give feedback.)
c. Work problem A. Put your pencil down when
  you know how many nails the man ended
  up with.
  (Observe children and give feedback.)
  (Teacher reference:)

- Read the problem and the answer you wrote
  for A. Get ready. (Signal.) *62 + 14 = 76.*
- How many nails did the man end up with?
  (Signal.) *76.*

from Lesson 44, Exercise 7

On the following lessons, the amount of structure
the teacher provides diminishes and some of the
problems have 3-digit numbers.

Here's part of the exercise from Lesson 57:

EXERCISE 10: WORD PROBLEMS (COLUMNS) REMEDY

a. Find part 6 on worksheet 57. ✔
(Teacher reference:)   R Part A

You're going to write the symbols for word
problems in columns and work them. The
equals bars are already shown.
- Touch where you'll write the symbols for
  problem A. ✔
  Listen to problem A: 765 students were in
  school. 562 of those students left the school.
  How many students ended up in the school?
- Listen again and write the number for the first
  part: 765 students were in school. ✔
- Write the sign and the number for the next
  part: 562 of those students left the school. ✔
- Touch and read the symbols you wrote for
  problem A. Get ready. (Signal.) *765 minus
  562 (equals).*
b. Work problem A and figure out how many
  students ended up in the school.
  (Observe children and give feedback.)
- For problem A, you wrote 765 minus 562.
  What's the answer? (Signal.) *203.*
- So how many students ended up in the
  school? (Signal.) *203.*

from Lesson 57, Exercise 10

COMPARISON WORD PROBLEMS

Starting on Lesson 77, students work with pairs of lines, such as:

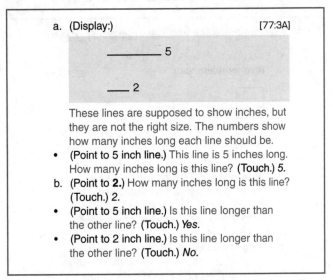

from Lesson 77, Exercise 3

Students say problems to figure out how much longer or shorter one of the lines is (9 – 3).

Students work comparison problems in their Workbook.

Here's the problem set from Lesson 81:

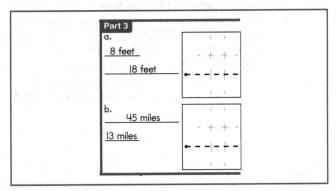

Workbook Lesson 81, Part 3

Starting on Lesson 83, students learn to make comparative statements. This is a preskill for problems that compare two items.

By Lesson 85, students have practiced saying sentences of the form "This line is ___ inches longer than the other line."

Here's the first part of the exercise from Lesson 85:

EXERCISE 3: COMPARISON STATEMENTS
LENGTH

a. (Display:)   [85:3A]

> 8
> 3

These lines are supposed to show inches.
- Which line is longer? (Signal.) *The top line.*
- Which line is shorter? (Signal.) *The bottom line.*
- Raise your hand when you know how much longer the top line is. ✔
- How much longer is the top line? (Signal.) *5 inches.*
- Say the sentence about the top line. (Signal.) *The top line is 5 inches longer than the bottom line.*
- Say the sentence about the bottom line. (Signal.) *The bottom line is 5 inches shorter than the top line.*
- (Repeat until firm.)

b. (Display:)   [85:3B]

> I
> 9

These lines are supposed to show inches.
- Which line is longer? (Signal.) *The bottom line.*
- Which line is shorter? (Signal.) *The top line.*
- Raise your hand when you know how much longer the bottom line is. ✔
- How much longer is the bottom line? (Signal.) *8 inches.*
- Say the sentence about the bottom line. (Signal.) *The bottom line is 8 inches longer than the top line.*
- Say the sentence about the top line. (Signal.) *The top line is 8 inches shorter than the bottom line.*
- (Repeat until firm.)

---

from Lesson 85, Exercise 3

Teaching Note: These statements are difficult for some students to say. Often students have trouble with the words that tell how much more or less (*five inches shorter than; five inches longer than*).

An effective correction is to model the sentence two times before directing the students to say it.

- Listen: The bottom line is 5 inches shorter than the top line.
- The bottom line is 5 inches shorter than the top line.
- Say the sentence about the bottom line.

On following lessons, students practice saying comparative statements about different attributes: how old the dogs are, how much the dogs weigh, how tall the trees are.

Here's part of the exercise from Lesson 88:

EXERCISE 2: COMPARISON STATEMENTS `REMEDY`

a. (Display:) [88:2A]

The number for each dog shows how many years old that dog is.
- (Point to black dog.) My turn: How old is the black dog? 8 years.
- (Point to white dog.) How old is the white dog? 5 years.
- Your turn: Which dog is older, the black dog or the white dog? (Signal.) *The black dog.*
- Say the problem for figuring out how much older the black dog is. (Signal.) *8 minus 5.*
- Say the problem for figuring how much younger the white dog is. (Signal.) *8 minus 5.*
- Tell me how much older the black dog is. Get ready. (Signal.) *3 years.*
- Say the sentence about the black dog. (Signal.) *The black dog is 3 years older than the white dog.*
- Say the sentence about the white dog. (Signal.) *The white dog is 3 years younger than the black dog.*
  (Repeat until firm.)

---

b. (Display:) [88:2B]

The number for each person tells how many years old that person is.
- (Point to boy.) How old is the boy? (Signal.) *10 years.*
- (Point to girl.) How old is the girl? (Signal.) *19 years.*
- Say the problem for figuring out how much younger the boy is. (Signal.) *19 minus 10.*
- Say the problem for figuring how much older the girl is. (Signal.) *19 minus 10.*
- Tell me how much younger the boy is. Get ready. (Signal.) *9 years.*
- Say the sentence about the boy. (Signal.) *The boy is 9 years younger than the girl.*
- Say the sentence about the girl. (Signal.) *The girl is 9 years older than the boy.*
  (Repeat until firm.)

from Lesson 88, Exercise 2

On Lesson 92, students work problems in which the numbers for two objects are given. Students figure out how much more or less one of the objects is.

Here's part of the exercise from Lesson 92:

EXERCISE 7: WORD PROBLEMS
COMPARISONS

a. Find part 3. ✔
(Teacher reference.)

For problem A, I'm going to tell you how heavy two dogs are. You're going to write the problem and figure out how much heavier one dog is than the other dog.
- Listen to problem A: A poodle weighs 21 pounds. A Labrador weighs 134 pounds.
- Listen to problem A again and answer some questions. A poodle weighs 21 pounds. How heavy is the poodle? (Signal.) *21 pounds.*
- Listen: A Labrador weighs 134 pounds. How heavy is the Labrador? (Signal.) *134 pounds.*
- Which dog is heavier, the poodle or the Labrador? (Signal.) *The Labrador.*
- Say the problem for figuring out how much heavier the Labrador is than the poodle. (Signal.) *134 minus 21.*
- Write the problem for A and work it. Put your pencil down when you know how many pounds heavier the Labrador is than the poodle. **(Observe children and give feedback.)**
b. Check your work.
- Read the problem and the answer for A. Get ready. (Signal.) *134 – 21 = 113.*
- The Labrador is how much heavier than the poodle? (Signal.) *113 pounds.*

from Lesson 92, Exercise 7

Students work similar problems on the following lessons. Starting on Lesson 105, students work mixed sets of problems. Some are action problems, some are class problems, and some involve comparisons.

Fractions

Students learn to write fractions that describe shapes that are divided into equal parts.

Starting on Lesson 117, students learn that the bottom number of the fraction tells the number of equal parts and that if a shape has 7 equal parts, each part is called a seventh. If a shape has 4 equal parts, each part is called a fourth.

Here's part of the exercise from Lesson 117:

EXERCISE 7: FRACTIONS   REMEDY

a. (Display:)   [117:7A]

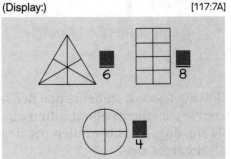

- (Point to triangle.) What shape is this? (Touch.) *(A) triangle.*
  This triangle is divided into parts that are the same size.
- (Point to **6.**) This number shows how many parts the triangle is divided into. How many parts? (Touch.) *6.*
- Yes, it is divided into 6 parts. Each part is called a sixth. Say **sixth.** (Signal.) *Sixth.*
- What is each part called? (Signal.) *(A) sixth.*
  (Repeat until firm.)
b. (Point to rectangle.) What shape is this? (Touch.) *(A) rectangle.*
- (Point to **8.**) How many equal parts is this rectangle divided into? (Touch.) *8.*
- So each part is called **an eighth.** What is each part called? (Signal.) *(An) eighth.*
c. (Point to circle.) What shape is this? (Touch.) *(A) circle.*
- (Point to **4.**) How many equal parts is this circle divided into? (Touch.) *4.*
- What is each part called? (Signal.) *(A) fourth.*
  (Repeat steps b and c until firm.)
d. Look at the triangle again and get ready to tell me what each part is called. ✔
- What is a part of the triangle called? (Signal.) *(A) sixth.*
e. Look at the rectangle and get ready to tell me what each part is called. ✔
- What is a part of the rectangle called? (Signal.) *(An) eighth.*
f. Look at the circle and get ready to tell me what each part is called. ✔
- What is a part of the circle called? (Signal.) *(A) fourth.*
  (Repeat steps d through f until firm.)

from Lesson 117, Exercise 7

Connecting Math Concepts

Halves are introduced on Lesson 119. This fractional part is delayed because it is irregular. For the others, the number of parts prompts the name of the parts. Halves are irregular because the name for two equal parts is not *twoths*.

Here's part of the exercise from Lesson 119:

c. The names for some parts are tricky because they don't tell how many parts the shape is divided into.
  • Listen: Fifths. What name? (Signal.) *Fifths.*
  • Who thinks they know how many equal parts a shape with fifths is divided into? (**Call on a student.**) *5.*
    Yes, if the parts of a shape are fifths, the shape is divided into 5 parts.
d. Again: Fifths. What name? (Signal.) *Fifths.*
  • A shape with fifths is divided into how many parts? (Signal.) *5.*
e. Here's another tricky name: Halves. What name? (Signal.) *Halves.*
  • A shape with halves is divided into 2 equal parts. A shape with halves is divided into how many parts? (Signal.) *2.*
f. Let's do those again.
  • Listen: Fifths. What name? (Signal.) *Fifths.*
  • A shape with fifths is divided into how many equal parts? (Signal.) *5.*
g. Listen: Halves. What name? (Signal.) *Halves.*
  • A shape with halves is divided into how many equal parts? (Signal.) *2.*
  (Repeat steps f and g until firm.)
h. Listen: Sevenths. What name? (Signal.) *Sevenths.*
  • A shape with sevenths is divided into how many equal parts? (Signal.) *7.*
i. (Repeat the following tasks for the following names:)

Listen: __. What name?		A shape with __ is divided into how many equal parts?	
Halves	*Halves*	Halves	2
Tenths	*Tenths*	Tenths	10
Fifths	*Fifths*	Fifths	5

(Repeat names that were not firm.)

from Lesson 119, Exercise 2

Students learn about the top number of fractions on Lesson 121.

On Lesson 122, students write fractions from pictures.

Here's that part of the exercise:

f. Find part 2 on your worksheet. ✔
  (Teacher reference:)

  You're going to write the number of equal parts for each shape and how many of those parts are shaded. Then you'll tell me the fraction.
  • Count the equal parts for shape A and write it in the space below the bar.
    (Observe children and give feedback.)
  • Everybody, how many equal parts is shape A divided into? (Signal.) *4.*
  • What are the parts of shape A called? (Signal.) *Fourths.*
  (Display:)   [122:5H]

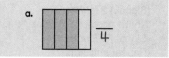

  Here's what you should have.
g. Count the parts that are shaded for shape A and write that number above the bar.
  (Observe children and give feedback.)
  • Shape A: How many fourths are shaded? (Signal.) *3.*
  • So what's the fraction for shape A? (Signal.) *3 fourths.*
  (Add to show:)   [122:5I]

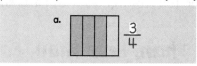

  Here's the fraction you should have for shape A.
h. Write the equal parts for shape B below the bar.
  (Observe children and give feedback.)
  • Everybody, how many equal parts is shape B divided into? (Signal.) *2.*
  • What are the parts of shape B called? (Signal.) *Halves.*
i. Write the number of parts that are shaded for shape B above the bar.
  (Observe children and give feedback.)
  • Shape B: How many halves are shaded? (Signal.) *2.*
  • So what's the fraction for shape B? (Signal.) *2 halves.*
  (Display:)   [122:5J]

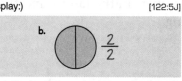

  Here's the fraction you should have for shape B.
j. Write the equal parts for shape C below the bar.
  (Observe children and give feedback.)
  • Everybody, how many equal parts is shape C divided into? (Signal.) *6.*
  • What are the parts of shape C called? (Signal.) *Sixths.*
k. Write the number of parts that are shaded for shape C above the bar.
  (Observe children and give feedback.)
  • Shape C: How many sixths are shaded? (Signal.) *1.*

from Lesson 122, Exercise 5

On Lesson 123, students learn that another name for fourths is *quarters*.

Here's part of the exercise from Lesson 123:

---

c. Now I'm going to teach you another name for fourths.
- Another way to say fourths is **quarters.** What's another way to say fourths? (Signal.) *Quarters.*
- What's another way to say quarters? (Signal.) *Fourths.*
  (Repeat until firm.)
d. Another way to say 3 fourths is 3 quarters.
- What's another way to say 3 fourths? (Signal.) *3 quarters.*
- What's another way to say 1 fourth? (Signal.) *1 quarter.*
- What's another way to say 2 quarters? (Signal.) *2 fourths.*
- What's another way to say 4 quarters? (Signal.) *4 fourths.*
  (Repeat until firm.)
- You can remember the new name for fourths, because 1 dollar equals 4 quarters. How many quarters equal a dollar? (Signal.) *4.*

from Lesson 123, Exercise 6

On the following lessons, students continue to write fractions for figures with parts shaded.

More Than, Less Than, Equals

This track teaches students the signs >, <, =, how to interpret them, and how to use them in statements that show number relationships.

The track begins on Lesson 81. At this point in the program, students have a good understanding of more and less. They know numbers that are more or less than specified numbers. They also know number family relationships, especially the notion that the big number is greater than the other numbers in the family and that it is reached by adding.

COMPLETING STATEMENTS

On Lesson 81, students learn the signs >, <. The teaching points out that each sign has a big end and a small end. The small end is the point. The numbers on either side of the sign are related to the ends of the sign. The number next to the small end of the sign is smaller than the number next to the big end of the sign.

Here's the part from Lesson 81:

EXERCISE 4: GREATER THAN/LESS THAN SIGN | REMEDY

a. (Display:) W   [81:4A]

> <

These are two new signs.
- (Point to left side of >.) This sign has a bigger end (touch) and a (touch right side of >) smaller end.
- (Point to right side of <.) This sign has a bigger end (touch) and a (touch left side of <) smaller end.
b. Tell me if the end I touch of each sign is bigger or smaller.
- (Point to left side of <.) Which end? (Touch.) *Smaller.*
- (Point to right side of <.) Which end? (Touch.) *Bigger.*
c. (Point to left side of >.) Which end? (Touch.) *Bigger.*
- (Point to right side of >.) Which end? (Touch.) *Smaller.*
  (Repeat steps b and c until firm.)
d. We're going to use these signs to tell if numbers are bigger or smaller.
  Here's how the signs work: The bigger number is next to the end that's bigger. The smaller number is next to the end that's smaller.
- Which end is next to the bigger number, the end that's bigger or smaller? (Signal.) *Bigger.*
- Which end is next to the smaller number, the end that's bigger or smaller? (Signal.) *Smaller.*
e. (Display:) W   [81:4B]

6   3

8   9

7   5

We're going to write one of the signs between the numbers in each row.
- (Point to **6 3**.) Read these numbers. (Touch.) *6 (and) 3.*
- Which number is bigger, 6 or 3? (Signal.) *6.*
  So I make the sign so the bigger end is next to the 6.
  (Add to show:)   [81:4C]

6 > 3

---

f. Again: Tell me which number is bigger, 6 or 3. (Signal.) *6.*
- Is the bigger end of the sign next to the bigger number? (Signal.) *Yes.*
- Is the smaller end of the sign next to the smaller number? (Signal.) *Yes.*
  So this is the right sign.
g. (Point to **8   9**.) Read these numbers. (Touch.) *8 (and) 9.*
- Which number is bigger, 8 or 9? (Signal.) *9.*
- So I make the sign with the bigger end next to which number? (Signal.) *9.*
  (Add to show:)   [81:4D]

  8 < 9

h. Again: Tell me which number is bigger, 8 or 9. (Signal.) *9.*
- Is the bigger end of the sign next to the bigger number? (Signal.) *Yes.*
- Is the smaller end of the sign next to the smaller number? (Signal.) *Yes.*
- So is this sign right? (Signal.) *Yes.*
i. (Point to **7   5**.) Read these numbers. (Touch.) *7 (and) 5.*
- Which number is bigger, 7 or 5? (Signal.) *7.*
- So does the end that's bigger or smaller go next to the 7? (Signal.) *Bigger.*
- Which end goes next to the 5? (Signal.) *(The) smaller (end).*
  (Add to show:)   [81:4E]

  7 < 5

- Is the bigger end of the sign next to the bigger number? (Signal.) *No.*
- So is this sign right? (Signal.) *No.*
  (Change to show:)   [81:4F]

  7 > 5

- Is the bigger end of the sign next to the bigger number? (Signal.) *Yes.*
- Is the smaller end of the sign next to the smaller number? (Signal.) *Yes.*
- So is this sign right? (Signal.) *Yes.*

from Lesson 81, Exercise 4

Teaching Note: Make sure you repeat steps B and C until students are very firm. This will reduce possible problems they may have later. It will also give you practice teaching the discrimination some students have trouble learning.

A > B

A is the bigger end of the sign. B is the smaller end of the sign.

8 < 9

8 is next to the smaller end of the sign. 9 is next to the bigger end of the sign.

If students have trouble, start by touching the bigger end of the sign. Ask,

- Is this the bigger end of the sign next to the bigger number?
- So this is the right sign.

If students can't see the basis for calling one of the ends bigger, point to the right side.

- Look how far apart the lines are on this end. So this end is bigger.
- (Point to smaller end.) Here the lines are together. So this end is smaller.

Practice correcting possible mistakes before presenting this exercise. Remember some students (and teachers) don't see the basis for one end of the sign being bigger than the other end.

On later lessons students work on a mix of problems that include those that have equal sides.

Here's the set of problems from Lesson 88:

Part 4

a. 46   53
b. 16   16
c. 57   48
d. 24   24
e. 117   130
f. 43   35

Workbook Lesson 88, Part 4

Students write the missing signs. Note that the equal sign has ends that are the same size, so it is consistent with the other two signs.

ASSOCIATIVE PROPERTY

Teaching the associative property of addition begins on Lesson 121. This work involves equations only. The problems start out with added values on either side of the equation. For example:

$$10 + 6 = 10 + 3 + 3$$

To show that the sides are equal, we group $3 + 3$ as $(3 + 3)$. The values equal 6, so the sides of the equation are equal.

Here's part of the exercise from Lesson 121:

EXERCISE 2: ASSOCIATIVE PROPERTY
EQUALITY

a. (Display:)   [121:2A]

    $10 + 6 = 10 + 3 + 3$ $47 + 28 = 40 + 7 + 28$

    $50 + 10 = 50 + 9 + 1$ $54 + 10 = 44 + 10 + 10$

All of these equations have numbers that are added on both sides. All of them are equal and I'll show you how to tell.
- (Point to **10 + 6**.) Read this equation. (Signal.) *10 + 6 = 10 + 3 + 3.*
- (Point to **10 + 6**.) What is 10 added to on this side of the equation? (Signal.) *6.*
- (Point to **10 + 3**.) 10 is added to 3 plus 3 on this side of the equation. What is 10 added to on this side of the equation? (Signal.) *3 plus 3.*
- What is 3 plus 3? (Signal.) *6.*
(Add to show:)   [121:2B]

    $10 + 6 = 10 + \overset{6}{(3 + 3)}$ $47 + 28 = 40 + 7 + 28$

    $50 + 10 = 50 + 9 + 1$ $54 + 10 = 44 + 10 + 10$

- Are both sides equal? (Signal.) *Yes.* Both sides show 10 plus 6.

b. (Point to **50 + 10**.) Read this equation. (Signal.) *50 + 10 = 50 + 9 + 1.*
- (Point to **50 + 10**.) What is 50 added to on this side of the equation? (Signal.) *10.*
- (Point to **50 + 9**.) 50 is added to 9 plus 1 on this side of the equation. What is 50 added to on this side of the equation? (Signal.) *9 plus 1.*
- What is 9 plus 1? (Signal.) *10.*
(Add to show:)   [121:2C]

    $10 + 6 = 10 + \overset{6}{(3 + 3)}$ $47 + 28 = 40 + 7 + 28$

    $50 + 10 = 50 + \overset{10}{(9 + 1)}$ $54 + 10 = 44 + 10 + 10$

- Are both sides equal? (Signal.) *Yes.* Both sides show 50 plus 10.

---

from Lesson 121, Exercise 2

Students learn the name *parentheses,* and they learn to put parentheses around two added values in the equation. All problems have two numbers on one side and three numbers on the other side. Students first identify the number that is on both sides. Then they make the parentheses around the other two numbers on the side that has three numbers.

For example:

$$10 + 8 + 9 = 17 + 9$$

Students find the number that is the same on both sides: 9. Then they put parentheses around $10 + 8$.

$$(10 + 8) + 9 = 17 + 9$$

Then they write a number that shows how many are inside the parentheses.

$$\overset{18}{(10 + 8)} + 9 = 17 + 9$$

Then they test to see if the statement is true or false. In this case, the statement is $18 + 9 = 17 + 9$. The statement is false.

Students make a line to change the = into ≠.

$$\overset{18}{(10 + 8)} + 9 \neq 17 + 9$$

Here's part of the exercise from Lesson 123:

EXERCISE 2: ASSOCIATIVE PROPERTY
INEQUALITY

a. (Display:)   [123:2A]

$$10 + 8 + 9 = 17 + 9 \qquad 56 + 10 = 50 + 6 + 10$$

$$25 + 4 = 25 + 3 + 2 \qquad 22 + 1 + 7 = 21 + 7$$

Some of these equations are not true. We'll figure out if each equation is true. Then we'll fix the sign for the equations that are false.

- (Point to **10 + 8.**) Read this equation. (Signal.) *10 + 8 + 9 = 17 + 9.*
- (Point to **10 + 8 + 9.**) What is 9 added to on this side of the equation? (Signal.) *10 plus 8.*
- (Point to **17.**) What is 9 added to on this side of the equation? (Signal.) *17.*
- Say the part I make parentheses around. Get ready. (Signal.) *10 + 8.*
- What's 10 plus 8? (Signal.) *18.*
  (Add to show:)   [123:2B]

$$\overset{18}{(10 + 8)} + 9 = 17 + 9 \qquad 56 + 10 = 50 + 6 + 10$$

$$25 + 4 = 25 + 3 + 2 \qquad 22 + 1 + 7 = 21 + 7$$

- Are both sides equal? (Signal.) *No.*
- So is this equation true or false? (Signal.) *False.*
  (Add to show:)   [123:2C]

$$\overset{18}{(10 + 8)} + 9 \neq 17 + 9 \qquad 56 + 10 = 50 + 6 + 10$$

$$25 + 4 = 25 + 3 + 2 \qquad 22 + 1 + 7 = 21 + 7$$

- Read the statement. (Signal.) *10 plus 8 plus 9 is not equal to 17 plus 9.*

---

from Lesson 123, Exercise 2

Teaching Note: On the following exercise in the lesson, students have the same problem with no sign between the sides:

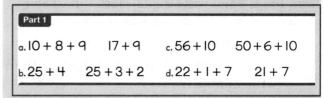

Workbook Lesson 123, Part 1

Students perform the same operations to determine whether the statement is true or false. Instead of writing the ≠ sign, they write the sign that shows which side is greater.

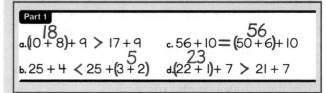

Answer Key Lesson 123, Part 1

Time

This track begins on Lesson 91 and continues through the end of the program. Students first learn to read and write digital clock time. Later students learn basic facts about analog clocks:

- They learn the direction hands rotate.
- They learn to identify the hour hand and the minute hand.
- They learn to interpret the time shown by each hand.
- They learn that the hour hand points to the numbers for hours.
- They learn that the minute hand shows how many times to count by 5.

   (If the minute hand points to 6, you count by five six times to find the number of minutes.)

- They learn to identify times for the hour or half hour by inspection.

DIGITAL CLOCKS

The following exercise from Lesson 92 is the second day students have worked with digital clocks.

EXERCISE 4: TELLING TIME
READING DIGITAL DISPLAY

a. (Display:)   [92:4A]

   Last time you learned how to read the time for clocks that showed zero minutes or 30 minutes.

- First you read the hour number. What do you read first? (Signal.) *The hour number.*
- Then, if the minute number is zero, you say o'clock. What do you say if the minute number is zero? (Signal.) *O'clock.*
- If the minute number is 30, you say 30. What do you say if the minute number is 30? (Signal.) *30.*

b. (Point to **9:00**.) Look at the minutes for this clock. Is the minute number zero? (Signal.) *Yes.*
- So do you say o'clock after the hour number? (Signal.) *Yes.*
- What do you say for the minutes? (Signal.) *O'clock.*
- Read the time for this clock. (Signal.) *9 o'clock.*

---

c. (Point to **2:30**.) Is the minute number for this clock zero? (Signal.) *No.*
- So do you say o'clock after the hour number? (Signal.) *No.*
- What do you say for the minutes? (Signal.) *30.*
- Read the time for this clock. (Signal.) *2 thirty.*

d. (Repeat the following tasks with remaining clocks:)

(Point to __.) Is the minute number for this clock zero?	So do you say o'clock after the hour number?	What do you say for the minutes?	Read the time for this clock.
11:00	Yes	Yes	O'clock	11 o'clock
10:30	No	No	Thirty	10 thirty
1:30	No	No	Thirty	1 thirty
5:00	Yes	Yes	O'clock	5 o'clock

(Repeat for clocks that were not firm.)
e. Read the time for these clocks again.
- (Point to **9:00**.) What time? (Touch.) *9 o'clock.*
- (Repeat for remaining clocks: 2 thirty, 11 o'clock, 10 thirty, 1 thirty, 5 o'clock.)

Lesson 92, Exercise 4

Teaching Note: There is a parallel between digital-time notation and money notation. For time, there is a colon after the hours. The minutes are always expressed as two digits. If the time is zero minutes after 3, the time is shown as: 3:00 and is read as **three o'clock.** This is parallel to the dollar amount, $3.00.

For 3:30, the time is read as **three thirty**, which is also parallel to the dollar-and-cent amount $3.30.

Following lessons review basic digital reading. By Lesson 94, students are well practiced in reading most digital times. On this lesson, they start writing digital times.

Here's the part of the exercise that introduces writing:

f. Find part 4 on your worksheet. ✔
    (Teacher reference:)   R Part A

    a. ___ : ___
    b. ___ : ___
    c. ___ : ___
    d. ___ : ___

    Now I'll say times and you'll write them. You'll write the hour number before the dots. You'll write the minute number after the dots.
    • If I say o'clock, what will you write after the dots? (Signal.) *Zero, zero.*
g. Touch where you'll write time A. ✔
    • Listen: 11 o'clock. What's time A? (Signal.) *11 o'clock.*
    • What's the hour number for 11 o'clock? (Signal.) *11.*
    • What's the number of minutes for 11 o'clock? (Signal.) *Zero.*
    • So what will you write after the dots? (Signal.) *Zero, zero.*
    • Write 11 o'clock in space A.
      (Observe children and give feedback.)
      (Display:)   [94:8B]

    a.   11:00

    Here's what you should have written for 11 o'clock.
h. Time B is 2 fifteen. What's time B? (Signal.) *2 fifteen.*
    • What's the hour number for 2 fifteen? (Signal.) *2.*
    • What's the number of minutes for 2 fifteen? (Signal.) *15.*
    • Write 2 fifteen in space B.
      (Observe children and give feedback.)
      (Display:)   [94:8C]

    b.   2:15

    Here's what you should have written for 2 fifteen.

i. Time C is 1 o'clock. What's time C? (Signal.) *1 o'clock.*
    • What's the hour number for 1 o'clock? (Signal.) *1.*
    • What's the number of minutes for 1 o'clock? (Signal.) *Zero.*
    • Write 1 o'clock in space C. Remember what to write for the minutes.
      (Observe children and give feedback.)
      (Display:)   [94:8D]

    c.   1:00

    Here's what you should have written for 1 o'clock.
j. Time D is 12 thirty. What's time D? (Signal.) *12 thirty.*
    • What's the hour number for 12 thirty? (Signal.) *12.*
    • What's the number of minutes for 12 thirty? (Signal.) *30.*
    • Write 12 thirty in space D.
      (Observe children and give feedback.)
      (Display:)   [94:8E]

    d.   12:30

    Here's what you should have written for 12 thirty.
k. Now you'll read the times you wrote for part 4.
    • Read time A. (Signal.) *11 o'clock.*
    • (Repeat for:) B, *2 fifteen;* C, *1 o'clock;* D, *12 thirty.*

from Lesson 94, Exercise 8

On Lesson 96, students learn the convention for minutes like :05 or :09. They learn to read these times as **Oh five,** and **Oh nine.** Students learn this convention through writing times.

Here's the first part of the exercise from Lesson 96:

EXERCISE 7: TELLING TIME
WRITING 1-DIGIT NUMBERS   REMEDY

a. Find part 3. ✔
   (Teacher reference:)   R Part B

   a. __:__
   b. __:__
   c. __:__
   d. __:__
   e. __:__
   f. __:__

   I'll say a time for each space. You'll write
   the time. The dots for each time are already
   shown.
   • If I say o'clock, what will you write after the
     dots? (Signal.) *Zero, zero.*
b. Time A is 6 o'clock. What's time A? (Signal.)
   6 o'clock.
   • What do you write after the dots for 6 o'clock?
     (Signal.) *Zero, zero.*
   • Write 6 o'clock in space A.
     (Observe children and give feedback.)
     (Display:)   [96:7A]

   | a.   **6:00** |

   Here's what you should have written for
   6 o'clock.
c. Time B is 5 oh 8. What's time B? (Signal.) *5 oh 8.*
   • What's the hour number for 5 oh 8? (Signal.) *5.*
   • The minutes are **oh 8,** so you write zero, 8
     after the dots. What do you write after the
     dots? (Signal.) *Zero, 8.*
   • Write the time for 5 oh 8 in space B.
     (Observe children and give feedback.)
     (Display:)   [96:7B]

   | b.   **5:08** |

   Here's what you should have written for 5 oh 8.
d. Time C is 12 oh 4. What's time C? (Signal.)
   12 oh 4.
   • What do you write after the dots for 12 oh 4?
     (Signal.) *Zero, 4.*
   • Write the time for 12 oh 4 in space C.
     (Observe children and give feedback.)
     (Display:)   [96:7C]

   | c.   **12:04** |

   Here's what you should have written for 12 oh 4.

---

from Lesson 96, Exercise 7

Teaching Note: Students are led through time B (5:08), but there is little structure for the next time (12:04). Students should have no trouble with these items.

Also note that after students have written all the times, they read them.

ANALOG CLOCKS

Students learn about analog clocks starting on Lesson 101.

On this lesson, students learn the direction the hands rotate and learn that the longer hand is the minute hand.

EXERCISE 3: TELLING TIME
ANALOG CLOCK HANDS   REMEDY

a. (Display:)   [101:3A]

   You've read the times on clocks that show
   numbers for hours and minutes. These are
   clocks too. But to figure out the time these
   clocks show, you have to look at the hands on
   the clock.
   • (Point to the minute hand on 1st clock.)
     This hand is the minute hand. Which hand?
     (Touch.) *The minute hand.*
   • The minute hand is longer than the other
     hand. Which hand is longer? (Signal.) *The
     minute hand.*
   • (Repeat until firm.)
b. (Point to hour hand on second clock.) Is this
   hand longer than the other hand? (Signal.) *No.*
   • So is it the minute hand? (Signal.) *No.*
   • (Point to the minute hand on second clock.)
     Which hand is this? (Signal.) *The minute hand.*
c. (Point to the minute hand on second clock.)
   Is this hand longer than the other hand?
   (Signal.) *Yes.*
   • So is it the minute hand? (Signal.) *Yes.*
d. (Point to hour hand on first clock.) Is this hand
   longer than the other hand? (Signal.) *No.*
   • So is it the minute hand? (Signal.) *No.*
   • (Repeat steps b through d until firm.)

---

e. Before you can tell time on a clock with hands, you have to know which direction the hands move.
(Display:) [101:3B]

The arrows show the direction that the hands move.
- (Point in the same direction the minute hand arrow on the first clock is pointing.) The minute hand on this clock is moving this way. (Touch 3 on first clock.) This is the next number the minute hand will point to. What's the next number the minute hand will point to? (Signal.) *3.*
- (Point to **2** on first clock.) This is the last number the minute hand pointed to. What's the last number the minute hand pointed to? (Signal.) *2.*

from Lesson 101, Exercise 3

Teaching Note: The direction the hands move is particularly important for the hour hand (which is taught several lessons later). The number for the hour is the last number the hour hand pointed to. If you know the direction the hands move, you can find the last number the hand pointed to.

The hour hand is not taught until Lesson 104. In the meantime, students practice drawing arrows to show the direction the hands rotate and identifying the last number the hand pointed to.

Students learn to count for the minute hand on Lesson 108.

Here's part of the exercise:

i. (Display:) [108:5C]

a.

5:

Here's what you should have for clock A. I'll show you how to count minutes.
- (Point to minute marks between numbers.) It takes 1 minute for the minute hand to go from one of these marks to the next mark. It takes 5 minutes for the minute hand to go from one number to the next number.
- (Touch 12.) How long does it take the minute hand to go from here to (touch 1) here? (Signal.) *5 minutes.*
- (Touch 1.) How long does it take the minute hand to go from here to (touch 2) here? (Signal.) *5 minutes.*
(Repeat for next numbers until firm.)

j. I'll count the minutes for clock A the fast way. After I get to the minute hand, tell me to stop. (Touch 1 through 11 as you count.) 5, 10, 15, 20, 25, 30, 35, 40, 45, 50, 55. *Stop.*
- Your turn: Touch 12 on clock A. ✔
- You'll touch the numbers and count for the minutes. Stop after you get to the minute hand. Get ready. (Tap 11.) *5, 10, 15, 20, 25, 30, 35, 40, 45, 50, 55.*
(Repeat until firm.)
- What's the minute number for clock A? (Signal.) *55.*

k. Touch 12 on clock B. Touch the numbers and count for the minutes. Stop after you get to the minute hand. Get ready. (Tap 5.) *5, 10, 15, 20, 25.*
(Repeat until firm.)
- What's the minute number for clock B? (Signal.) *25.*

l. Touch 12 on clock C. Touch the numbers and count for the minutes. Stop after you get to the minute hand. Get ready. (Tap 8.) *5, 10, 15, 20, 25, 30, 35, 40.*
(Repeat until firm.)
- What's the minute number for clock C? (Signal.) *40.*

m. Touch 12 on clock D. Touch the numbers and count for the minutes. Stop after you get to the minute hand. Get ready. (Tap 2.) *5, 10.*
(Repeat until firm.)
- What's the minute number for clock D? (Signal.) *10.*

n. Touch 12 on clock A again. Touch the numbers and count for the minutes. Stop after you get to the minute hand. Get ready. (Tap 11.) *5, 10, 15, 20, 25, 30, 35, 40, 45, 50, 55.*
- What's the minute number for clock A? (Signal.) *55.*
(Repeat steps j through n that were not firm.)

from Lesson 108, Exercise 5

Students practice writing times in structured exercises. They draw arrows for the hands. They write the number for hours, then the number for minutes.

Here's an exercise in which students take all the steps with a minimum of structure.

EXERCISE 5: TELLING TIME
ANALOG HOURS AND MINUTES

a. (Distribute unopened workbooks to children.)
• Open your workbook to Lesson 112 and find part 1.
   (Observe children and give feedback.)
   (Teacher reference:)

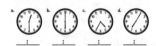

For these clocks you're going to make an arrow to show the direction the hour hand is moving. Then you'll write the time for each clock.

• For each clock, make an arrow to show the direction the hour hand is moving. Then write the hour number for each clock. Put your pencil down when you've made arrows for all of the hour hands and written the hour number for each clock.
   (Observe children and give feedback.)
b. Check your work. You'll read the hour number for each clock.
┌ • Clock A. (Signal.) *12.*
└ • (Repeat for:) B, *6;* C, *4;* D, *7.*
   (Display:)   [112:5A]

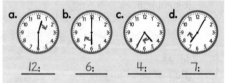

Here's what you should have for the clocks in part 1.
c. Now you'll figure out the minute number for each clock.
• How many minutes do you count for each number on a clock? (Signal.) *5.*
• Touch each number on clock A and count to yourself. Raise your hand when you know the minute number for clock A.
   (Observe children and give feedback.)
• What's the minute number for clock A? (Signal.) *30.*
d. Look at clock B and get ready to tell me the minute number.
• What's the minute number for clock B? (Signal.) *Zero.*

---

e. Touch each number on clock C and count to yourself. Raise your hand when you know the minute number for clock C.
   (Observe children and give feedback.)
• What's the minute number for clock C? (Signal.) *35.*
f. Look at clock D and get ready to tell me the minute number.
• What's the minute number for clock D? (Signal.) *5.*
g. Write the minutes for the clocks in part 1.
   (Observe children and give feedback.)
h. Check your work. You'll read the time for each clock.
┌ • Clock A. (Signal.) *Twelve 30.*
└ • (Repeat for:) B, *6 o'clock;* C, *Four 35;* D, *Seven oh 5.*
   (Display:)   [112:5B]

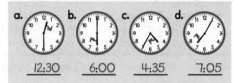

Here's what you should have for the clocks.
┌ i. Listen: When the minute hand is straight up, how many minutes does the clock show? (Signal.) *Zero.*
│ • When the minute hand is straight down, how many minutes does the clock show? (Signal.) *30.*
└ • (Repeat step i until firm.)

Lesson 112, Exercise 5

Starting on Lesson 110, students learn to identify times that show **o'clock** or **30.**

Here's part of the time-telling exercise from Lesson 113:

EXERCISE 6: TELLING TIME
ANALOG HOURS AND HALF HOURS   **REMEDY**

a. Think about the minute hand on clocks.
  • Listen: When the minute hand is straight up, how many minutes does the clock show? (Signal.) *Zero.*
  • When the minute hand is straight down, how many minutes does the clock show? (Signal.) *30.*
  • For each number on a clock, how many minutes do you count? (Signal.) *5.*
  (Repeat until firm.)

b. (Display:)   [113:6A]

The minute hand for each of these clocks shows zero minutes or 30 minutes. You're going to tell me the hour number and the minute number for each clock. Then you'll tell me the time the clock shows.

  • (Point to **5:00**.) What's the hour number for this clock? (Signal.) *5.*
  • What's the minute number? (Signal.) *Zero.*
  • Say the time this clock shows. (Signal.) *5 o'clock.*

c. (Repeat the following tasks for the remaining clocks:)

(Point to __.) What's the hour number for this clock?	What's the minute number?	Say the time this clock shows.
3:30	3	*Thirty*	*3:30*
10:30	10	*Thirty*	*10:30*
8:00	8	*Zero*	*8 o'clock*
7:30	7	*Thirty*	*7:30*
12:00	12	*Zero*	*12 o'clock*

(Repeat clocks that were not firm.)

---

from Lesson 113, Exercise 6

By Lesson 119, students tell times without making arrows or identifying hands. The sets of examples include times for the hour and half hour as well as other times.

Here's the exercise from Lesson 119:

EXERCISE 8: TELLING TIME
ANALOG HOURS AND HALF HOURS

a. Turn to the other side of worksheet 119 and find part 5. ✔
  (Teacher reference:)

  You're going to write the time for each clock. One clock shows zero minutes. Another clock shows 30 minutes. You'll count the minutes to yourself for the other two clocks.
  • Write the time for each clock. Put your pencil down when you've written the hour number and the minute number for each clock in part 5. (Observe children and give feedback.)

b. Check your work. You'll read the time for each clock.
  • Clock A. (Signal.) *Nine 35.*
  • (Repeat for:) B, *3 o'clock*; C, *Six 15*; D, *Eleven 30.*
  (Display:)   [119:8A]

Here's what you should have for the clocks.

Lesson 119, Exercise 8

Money

This track starts on Lesson 41 and continues through the end of the program. Students review some of the count-on skills they learned in *CMC Level A* for counting mixed sets of coins:

For this set, students start with 25 for the quarter and then count on for the dimes (35, 45) and for the pennies (46, 47). They also identify bills.

Students work with mixed sets of bills and coins. For example:

Students work problems that add or subtract dollar-and-cent amounts. For example:

$$\begin{array}{r} \$15.10 \\ +\ 5.68 \\ \hline \end{array}$$

Here's part of a review exercise from Lesson 48:

EXERCISE 7: COINS REMEDY
a. (Display:) [48:7A]

- (Point to dimes.) This row has dimes and pennies. My turn to count for the dimes, then count for each penny. (Touch dimes.) 10, 20, 30. (Keep your finger on last dime.) Now I count for each penny. Thirtyyy, (Touch pennies.) 31, 32, 33, 34.
- (Point to dimes.) Your turn: Count for the dimes. Then count for each penny. Get ready. (Touch dimes.) *10, 20, 30.* (Keep your finger on last dime.)
- How many cents do you have? (Signal.) *30.*
- Get 30 going and count for each penny. Get it going. *Thirtyyy.* (Touch pennies.) *31, 32, 33, 34.*

b. Find part 3 on worksheet 48. ✔ (Teacher reference:) R Part I

I'll count for the coins in row A. Touch each coin as I count.
- Finger over the first coin in row A. ✔
- (Children touch dimes as you count.) 10, 20, thirtyyy. (Children touch pennies as you count.) 31, 32, 33, 34.

c. Now you'll touch and count the coins in row A.
- Finger over the first coin in row A. ✔
- Count for the dimes. (Tap 3.) *10, 20, thirtyyy.* Count for the pennies. *(Tap 4.) 31, 32, 33, 34.* (Repeat until firm.)
- How many cents is row A worth? (Signal.) *34.*

from Lesson 48, Exercise 7

Teaching Note: Make sure that students are touching and counting properly.

A good idea is to repeat step B several times. Pay close attention to the transition from dimes to pennies. Students are to touch the last dime as long as you hold the number *thirtyyy*. They are to touch the first penny when you say 31.

On Lesson 55, students apply counting by fives to figure out how much groups of nickels are worth. Here's part of the exercise:

EXERCISE 9: COINS
NICKELS   **REMEDY**

a. (Display:)   [55:9A]

- (Point to nickels.) These coins are **nickels**. What are they? (Touch.) *Nickels.*
- A nickel is worth 5 cents. How much is a nickel worth? (Touch.) *5 cents.*
- So when you count nickels, what do you count by? (Signal.) *Fives.*
- (Point to 1st row.) Count by fives and figure out how much this row is worth. Get ready. (Touch.) *5, 10, 15, 20, 25, 30.*
- How much is this row worth? (Signal.) *30 cents.*
b. (Point to 2nd row.) You'll count and figure out how much this row is worth. Get ready. (Touch.) *5, 10, 15, 20, 25, 30, 35, 40.*
- How much is this row worth? (Signal.) *40 cents.*

---

from Lesson 55, Exercise 9

On Lesson 57, students count on for nickels. Here's part of the exercise:

EXERCISE 6: COINS

a. Find part 2 on worksheet 57. ✔
(Teacher reference:)

Group A has dimes and nickels.
- What do you count by for each dime? (Signal.) *Ten.*
- What do you count by for each nickel? (Signal.) *Five.*
b. I'll count. You'll touch the coins.
- Fingers ready. ✔
- (Children touch dimes.) 10, twentyyyy. (Children touch nickels.) 25, 30, 35, 40.
c. Your turn to touch and count.
- Fingers ready. ✔
- Touch and count for the dimes. (Signal.) *10, twentyyy.* Count for the nickels. (Signal.) *25, 30, 35, 40.*
- (Repeat until firm.)
- How many cents is group A worth? (Signal.) *40.*

---

from Lesson 57, Exercise 6

Connecting Math Concepts

On Lesson 61, students count on by fives for bills. Here's the set of items from Lesson 61:

---

from Lesson 61, Exercise 4

Also on Lesson 61, students count by 25 for quarters. Here's the first part of the introduction:

EXERCISE 6: COINS
QUARTERS   **REMEDY**

a. (Display:)   [61:6A]

- (Point to quarter.) This coin is a **quarter**. What coin is this? (Touch.) *(A) quarter.*
- A quarter is worth 25 cents. What's a quarter worth? (Signal.) *25 cents.*
(Repeat until firm.)
b. Listen: Here are the numbers for counting the cents for quarters: 25, 50, 75, 100.
- Listen again: 25, 50, 75, 100. Say the numbers for counting the cents for quarters. Get ready. (Signal.) *25, 50, 75, 100.*
c. Find part 2 on worksheet 61. ✔
(Teacher reference:)   **R Part H**

There are quarters in each of these groups of coins. We'll count for each group to figure out how much each group is worth.
- Touch the coins in group A. ✔
I'll count. You'll touch the coins as I count.
- Fingers over the first quarter. ✔
- Get ready. (Children touch quarters.) 25, 50.
d. Your turn to count the cents in group A. Get ready. (Tap 2.) *25, 50.*
- How many cents is group A worth? (Signal.) *50.*
- Write an equals and the number of cents for group A. ✔
e. Touch the coins in group B. ✔
You'll touch and count for the coins in group B.
- Fingers over the first quarter. ✔
- Get ready. (Tap 4.) *25, 50, 75, 100.*
(Repeat until firm.)
- How many cents is group B worth? (Signal.) *100.*
- Yes, 100 cents. 100 cents is a dollar. How many cents is a dollar? (Signal.) *100.*
- Write an equals and the number of cents for group B. ✔

---

from Lesson 61, Exercise 6

On Lesson 65, students learn that 100 cents equals one dollar. Students confirm that a dollar is worth 100 cents and confirm that rows of coins that equal 100 cents are worth a dollar.

Here's the first part of the exercise from Lesson 65:

EXERCISE 1: COINS EQUAL TO A DOLLAR

a. Listen: A dollar is worth 100 cents. How many cents is a dollar worth? (Signal.) *100.*
 • Listen: A dollar is worth 100 cents. So how many cents is 2 dollars worth? (Signal.) *200.*
 • How many cents is 5 dollars worth? (Signal.) *500.*
 • How many cents is 9 dollars worth? (Signal.) *900.*
 • How many cents is 1 dollar worth? (Signal.) *100.*
 (Repeat step a until firm.)

b. (Display:) [65:1A]

 • (Point to [1].) This shows 100 cents in dollars.
 • (Point to ⊙⊙⊙⊙.) This shows 100 cents in quarters.
 • (Point to ⊙⊙⊙⊙⊙⊙⊙⊙⊙⊙.) This shows 100 cents in dimes.
 • (Point to ⊙⊙⊙⊙⊙⊙⊙⊙⊙⊙.) This shows 100 cents in nickels.
 • (Point to [1].) How many cents is a dollar? (Signal.) *100.*
 So this row shows 100 cents.

c. (Point to ⊙.) How many cents is a quarter? (Signal.) *25.*
 • Count by 25s for the quarters. Get ready. (Touch.) *25, 50, 75, 100.*
 So we have 100 cents in quarters.

from Lesson 65, Exercise 1

On Lesson 68, students work with rows of coins. If a row has 100 cents, students circle the words one dollar. If the row does not have 100 cents, students write the number of cents for the row.

Here's the problem set from Lesson 68:

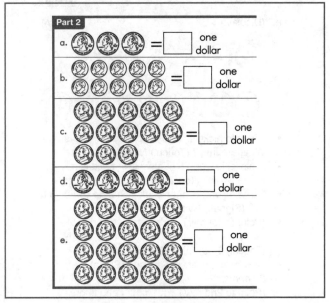

Workbook Lesson 68, Part 2

WRITING DOLLAR AMOUNTS

Starting on Lesson 70, students learn to write dollar-and-cent amounts. The first exercises have the dollar sign. Students write the number for dollars, then the decimal point, and then the number for cents.

Here's part of the exercise from Lesson 76. On the previous lesson, students learned to write the dollar sign.

EXERCISE 7: MONEY
WRITING DOLLAR AMOUNTS

REMEDY

a. Find part 3 on worksheet 76. ✔
   (Teacher reference:)

R Part C

   You're going to write dollars and cents.
- Touch space A. ✔
- You will write 5 dollars and 24 cents for space A. What will you write? (Signal.) *5 dollars and 24 cents.*
- Write the dollar sign for A. Write a capital S. Then make a straight line down through it. **(Observe children and give feedback.)**
- Now write the rest of the 5 dollars and 24 cents. Remember the dot after the number of dollars. **(Observe children and give feedback.)** Check your work.
- Touch the dollar sign you wrote. ✔
- Touch the number of dollars. ✔
- What number? (Signal.) *5.*
- Touch the dot. ✔
- Touch the number of cents. ✔
- How many cents? (Signal.) *24.*
b. Touch the space for B. ✔
- You'll write 9 dollars and 53 cents for space B. What will you write? (Signal.) *9 dollars and 53 cents.*
- Write the dollar sign. Remember, a capital S with a line straight down through it. ✔
- Now write the rest of 9 dollars and 53 cents. **(Observe children and give feedback.)**
- What's the first thing you wrote? (Signal.) *A dollar sign.*
- What's the next thing you wrote? (Signal.) *9.*
- What's the next thing you wrote? (Signal.) *(A) dot.*
- What's the last thing you wrote? (Signal.) *53.*

from Lesson 76, Exercise 7

Teaching Note: Observe students as they work the problems. Correct mistakes, particularly mistakes of omitting the dollar sign or the dot.

In step A, you present a workcheck that provides a check for the dollar sign, the number of dollars, the dot that follows the dollars, and the number of cents.

For most parts of this check, students touch symbols. Observe students closely because this check lists the details that students must attend to when writing dollar-and-cent amounts.

If they are weak at touching the correct symbols or in identifying the number of dollars and cents, repeat the workcheck. Praise students for correct responses.

COLUMN PROBLEMS

On Lesson 77, students work problems that add or subtract dollar-and-cent amounts. The routine students follow is designed to prompt their awareness that the numbers after the dot are cents and that the dollar sign is next to the dollar amounts.

The students first write the dot in the answer and work the problem for cents. Then students work the problem for dollars and write the dollar sign in the answer.

Here's the first Workbook problem:

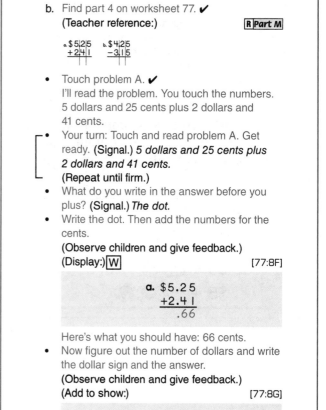

- Touch problem A. ✔
  I'll read the problem. You touch the numbers. 5 dollars and 25 cents plus 2 dollars and 41 cents.
- Your turn: Touch and read problem A. Get ready. (Signal.) *5 dollars and 25 cents plus 2 dollars and 41 cents.* (Repeat until firm.)
- What do you write in the answer before you plus? (Signal.) *The dot.*
- Write the dot. Then add the numbers for the cents. **(Observe children and give feedback.)** (Display:) W   [77:8F]

$$\begin{array}{r} \text{a. } \$5.25 \\ +2.41 \\ \hline .66 \end{array}$$

  Here's what you should have: 66 cents.
- Now figure out the number of dollars and write the dollar sign and the answer. **(Observe children and give feedback.)** (Add to show:)   [77:8G]

$$\begin{array}{r} \text{a. } \$5.25 \\ +2.41 \\ \hline \$7.66 \end{array}$$

  Here's the answer for the problem.
- Everybody, read the answer. (Signal.) *7 dollars and 66 cents.*

from Lesson 77, Exercise 8

On Lesson 80, students write dollar amounts that have fewer than ten cents. They learn that you must have two digits after the dot. For less than 10 cents, the first digit is zero.

Here's the review of the procedure that occurs on Lesson 81:

EXERCISE 2: DOLLARS WITH 1-DIGIT CENTS REMEDY

a. You learned about writing cents numbers that are less than 10. Remember, when you write cents that are less than 10, you write a zero before the number.

- If you want to write 3 cents, you write zero 3 after the dot.
- What do you write after the dot? (Signal.) *Zero 3.*
- If you want to write 4 cents, what do you write after the dot? (Signal.) *Zero 4.*
- If you want to write 2 cents, what do you write after the dot? (Signal.) *Zero 2.*

  (Repeat until firm.)

from Lesson 81, Exercise 2

On Lesson 90, students work problems that add cents to dollar amounts. For each item, students write the equal sign and the dollar-and-cent amount.

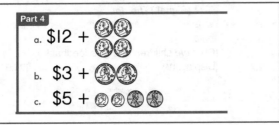

Workbook Lesson 90, Part 4

The last problem type introduced in the sequence first appears on Lesson 92. For each problem, students count bills and coins and complete the equations with the dollar-and-cent amounts.

Here's the problem set from Lesson 93:

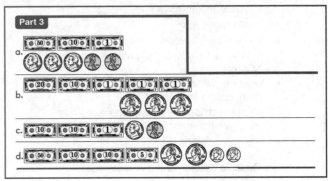

Workbook Lesson 93, Part 3

Geometry

The Geometry track starts on Lesson 61 and continues through the end of the program.

The first part of the Geometry track reviews two-dimensional forms students learned in *CMC Level A*: circle, triangle, rectangle, square, and hexagon. The track also reviews cubes and cylinders. The new two-dimensional shape *CMC Level B* presents is pentagon.

The new three-dimensional shapes are pyramids and prisms. The common features of pyramids are that they all have triangular sides, and all sides come to the same point.

Rectangular prisms are three-dimensional shapes that have six rectangular sides. A special case of a rectangular prism is a cube. All sides are identical squares. Other prisms have identical shapes on either end. The ends can be triangles, or hexagons, or any other straight-sided shape. The sides connecting the ends are rectangles.

Students also do tasks related to composition and decomposition of complex forms. By the end of the track, students have a good sense of shapes, three-dimensional objects, and their two-dimensional faces. The exercises near the end of the track provide the ultimate composition activities. Students identify the component faces, the number of each type, and they also identify a picture that shows the composed object.

TWO-DIMENSIONAL SHAPES

Students review two-dimensional shapes, starting on Lesson 61.

Here's the review exercise from Lesson 67:

EXERCISE 5: SHAPES
INTRODUCTION OF HEXAGON

a. (Display:)   [67:5A]

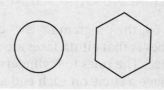

- (Point to ○.) What shape is this? (Touch.) *(A) circle.*
b. (Point to ⬡.) Is this shape a triangle? (Touch.) *No.*
- How many sides does a triangle have? (Signal.) *3.*
- Does this shape have more than 3 sides or less than 3 sides? (Touch.) *More than 3 sides.*
c. A rectangle has more than 3 sides. Is this shape a rectangle? (Touch.) *No.*
- How many sides does a rectangle have? (Signal.) *4.*
- (Point to ⬡.) Does this shape have more than 4 sides or less than 4 sides? (Touch.) *More than 4 sides.*
d. I'll touch. You'll count the sides of this shape. Get ready. (Touch.) *1, 2, 3, 4, 5, 6.*
- How many sides? (Touch.) *6.*
e. The name of this shape is hard to say. **Hexagon**. Listen again: **hexagon**.
- What's this shape? (Signal.) *Hexagon.*
- Yes, hexagons have 6 sides. Say the sentence. (Signal.) *Hexagons have 6 sides.*
f. Say the sentence about rectangles. (Signal.) *Rectangles have 4 sides.*
- Say the sentence about triangles. (Signal.) *Triangles have 3 sides.*
- Say the sentence about hexagons. (Signal.) *Hexagons have 6 sides.*
- (Repeat step f until firm.)
g. (Display:)   [67:5B]

These are street signs. You'll tell me the shape of each sign.
- (Point to yield.) What shape? (Touch.) *(A) triangle.*
- (Point to speed.) What shape? (Touch.) *(A) rectangle.*
- (Point to railroad crossing.) What shape? (Touch.) *(A) circle.*
- (Repeat until firm.)

Lesson 67, Exercise 5

Teaching Note: Students who have gone through *CMC Level A* should have no trouble with this exercise. Even though they are firm on the information, don't skip the exercise. Just go fast. Make sure that students are firm on all the items in step F.

THREE-DIMENSIONAL SHAPES

On Lesson 69, the program reviews cubes. The rule for cubes is that all six faces are squares of the same size. The rules for cylinders are that cylinders have a circle on each end and both circles are the same size.

The review of cylinders begins on Lesson 72. Here's part of the exercise.

---

b. (Display:)   [72:2B]

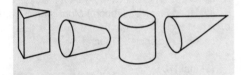

- (Point to ⬠.) Does this object have a circle on each end? (Touch.) *No.*
- So is this a cylinder? (Signal.) *No.*
c. (Point to ⬡.) Does this object have a circle on each end? (Touch.) *Yes.*
- Are the circles the same size? (Signal.) *No.*
- So is this a cylinder? (Signal.) *No.*
d. (Point to ⬡.) Does this object have a circle on each end? (Touch.) *Yes.*
- Are the circles the same size? (Signal.) *Yes.*
- So is this a cylinder? (Signal.) *Yes.*
e. (Point to ▷.) Does this object have a circle on each end? (Touch.) *No.*
- So is this a cylinder? (Signal.) *No.*
  Remember, a cylinder has a circle at each end, and the circles are the same size.
- What object has a circle at each end and the circles are the same size? (Signal.)
  (A) cylinder.

---

from Lesson 72, Exercise 2

On Lesson 100, pyramids are introduced. Students learn that pyramids may have bases of different shapes.

EXERCISE 7: 3-D OBJECTS
PYRAMIDS

a. Find part 3 on worksheet 100. ✔
    (Teacher reference:)

- The objects in part 3 are pyramids. Say **pyramid.** (Signal.) *Pyramid.*
- The side faces of pyramids are always triangles. What shape are the side faces? (Signal.) *Triangles.*
- But the bottom faces of pyramids can be a triangle, a rectangle, or a hexagon. Do the bottom faces of pyramids have to be triangles? (Signal.) *No.*
- Think about the top of pyramids. Do pyramids have a top face? (Signal.) *No.*

---

from Lesson 100, Exercise 7

Pentagon is introduced on Lesson 103. Students learn that a pentagon is a shape with five sides. They see that a pyramid may have a base that is a pentagon.

On Lesson 107, students identify the faces of three-dimensional objects. Then they indicate the picture that shows the shape that has this set of faces. Note that the set includes faces that cannot be observed in the pictures of the three-dimensional objects.

Here's part of the exercise from Lesson 107. At the beginning of this excerpt students have written letters on the faces.

(Teacher reference:)

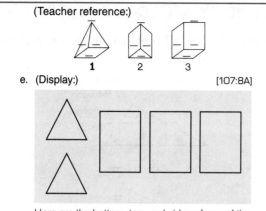

1   2   3

e. (Display:) [107:8A]

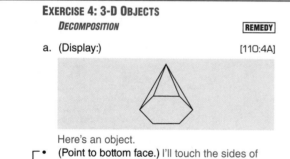

Here are the bottom, top, and sides of one of the objects. You're going to figure out which object.
- (Point to bottom face.) Here's the bottom face.
- Everybody, what is the shape of the bottom face? (Signal.) *(A) triangle.*
- (Point to top face.) Everybody, what is the shape of the top face? (Signal.) *(A) triangle.*
- (Point to side faces.) Here are the side faces. What shape are the side faces? (Touch.) *Rectangle(s).*
f. Circle the object that has the right top, bottom, and sides.
   (Observe children and give feedback.)
- Everybody, which object did you circle? (Signal.) *(Object) 2.*

from Lesson 107, Exercise 8

On Lesson 110, students work exercises in which they identify the number of different types of faces a three-dimensional shape has. Students do this by completing sets of equations that start with letters of shapes.

Here's part of the exercise from Lesson 110:

EXERCISE 4: 3-D OBJECTS
DECOMPOSITION   [REMEDY]

a. (Display:) [110:4A]

Here's an object.
- (Point to bottom face.) I'll touch the sides of the bottom face. Count them to yourself and figure out the shape of the bottom face and the number of side faces.
- (Touch each side of bottom face.) What's the shape of the bottom face? (Touch.) *(A) hexagon.*
- How many sides does the bottom face have? (Signal.) *6.*
- So how many side faces does this object have? (Signal.) *6.*
- What shape are the side faces? (Signal.) *Triangles(s).*
- Does this object have a top face? (Signal.) *No.*
   (Repeat until firm.)

---

b. (Distribute unopened workbooks to children.)
- Open your workbook to Lesson 110 and find part 1.
   (Observe children and give feedback.)
   (Teacher reference:) **R** **Test 12: Part G**

Part 1   Part 2
H =   H =
P =   P =
R =   R =
T =   T =

The letters in part 1 and part 2 stand for shapes.
- Touch H equals. ✔
   What shape does H stand for? (Signal.) *(A) hexagon.*
- Touch P equals. ✔
   What shape does P stand for? (Signal.) *(A) pentagon.*
- Touch R equals. ✔
   What shape does R stand for? (Signal.) *(A) rectangle.*
- Touch T equals. ✔
   What shape does T stand for? (Signal.) *(A) triangle.*
   (Repeat until firm.)
c. (Point to pyramid) Now you're going to complete each equation.
- How many faces of this pyramid are hexagons? (Touch.) *1.*
- Complete the equation for hexagons.
   (Observe children and give feedback.)
d. (Point to pyramid) How many faces are pentagons? (Touch.) *Zero.*
- How many faces are rectangles? (Touch.) *Zero.*
- Complete the equations for pentagons and rectangles.
   (Observe children and give feedback.)
e. (Point to pyramid) How many faces are triangles? (Touch.) *6.*
   Yes, there are 6 side faces that are triangles.
- Complete the equation for triangles.
   (Observe children and give feedback.)

from Lesson 110, Exercise 4

On Lesson 114, students learn to identify different types of prisms. They learn the rule for prisms: Prisms have side faces that are rectangles. They later learn the end faces may be squares, rectangles, triangles, pentagons, or hexagons.

Here's part of the exercises from Lesson 114:

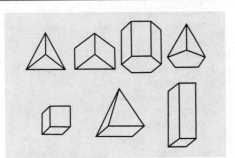

g. (Point at objects.) For these objects if it's not a pyramid, it's called a prism. Say **prism.** (Signal.) *Prism.*
• What are these objects that are not pyramids called? (Signal.) *Prisms.*
h. Prisms have side faces that are rectangles. What shape are the side faces of prisms? (Signal.) *Rectangles.*
• (Point to △.) Does this object have side faces that are rectangles? (Signal.) *No.*
• So is this a prism? (Signal.) *No.*
i. (Repeat the following tasks for the remaining objects:)

(Point to __.) Does this object have side faces that are rectangles?		So is this a prism?
Triangular prism	Yes	Yes
Hexagonal prism	Yes	Yes
Pentagonal pyramid	No	No
Cube	Yes	Yes
Rectangular pyramid	No	No
Rectangular prism	Yes	Yes

(Repeat for objects that were not firm.)
j. Now you'll tell me if each object is a pyramid or a prism.
• (Point to △.) Is this object a pyramid or a prism? (Touch.) *(A) pyramid.*
k. (Repeat the following tasks for the remaining objects:)

(Point to __.) Is this object a pyramid or a prism?	
Triangular prism	*(A) prism*
Hexagonal prism	*(A) prism*
Pentagonal pyramid	*(A) pyramid*
Cube	*(A) prism*
Rectangular pyramid	*(A) pyramid*
Rectangular prism	*(A) prism*

(Repeat for objects that were not firm.)
l. Look at the objects that are prisms and get ready to tell me how many of the prisms are cubes. ✔
• How many of these prisms are cubes? (Signal.) *1.*

from Lesson 114, Exercise 2

COMPOSITION/DECOMPOSITION

The work with two-dimensional figures presents items like this one:

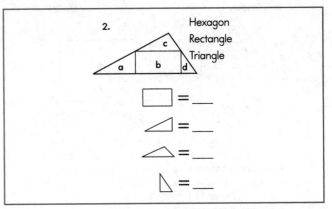

from Lesson 78, Exercise 5

Work with decomposition starts on Lesson 75. Students identify the whole shape and the component shapes.

Here's part of the exercise from Lesson 76:

c. (Distribute unopened workbooks to students.)
• Open your workbook to Lesson 76 and find part 1.
(Observe children and give feedback.)
(Teacher reference:) **R** Test 9: Part M

• Look at the whole figure. Is the shape for the whole figure a square or a hexagon? (Signal.) *A square.*
• Touch the first equals sign. ✔
You're going to write the letter for that part.
• What shape is that part? (Signal.) *(A) triangle.*
• Find the letter for the correct triangle and write the letter.
(Observe children and give feedback.)
• Everybody, what's the letter for the first equals? (Signal.) *D.*
d. Write letters for the rest of the parts.
(Observe children and give feedback.)
(Answer key:)

e. Check your work.
You wrote the letter D for the first equals.
• What letter did you write for the next equals? (Signal.) *B.*
• What letter did you write for the next equals? (Signal.) *C.*
• What letter did you write for the last equals? (Signal.) *A.*

from Lesson 76, Exercise 5

In step D, you observe and give students feedback. If students make mistakes, point out the relationship between the figure in the equation and the corresponding figure in the complex shape.

For example, a student misidentifies the first rectangle as A. Tell the student to touch the rectangle in the larger shape that looks just like the first rectangle. Then ask "What's the letter of the rectangle that looks just like the one in the equation?"

Students also work on composition exercises. For these the component parts are labeled. Students write the letters on the component parts of a complex form.

Here's part of the exercise from Lesson 87:

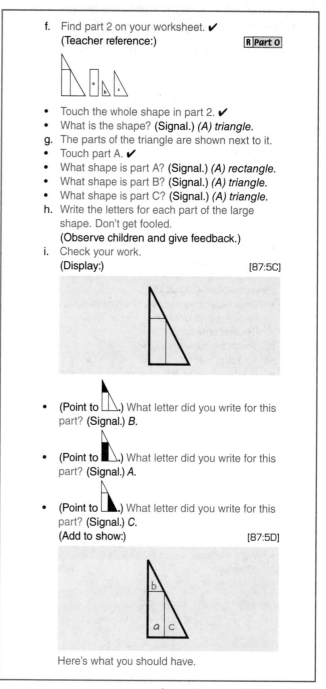

f. Find part 2 on your worksheet. ✔
(Teacher reference:) **R** Part O

• Touch the whole shape in part 2. ✔
• What is the shape? (Signal.) *(A) triangle.*
g. The parts of the triangle are shown next to it.
• Touch part A. ✔
• What shape is part A? (Signal.) *(A) rectangle.*
• What shape is part B? (Signal.) *(A) triangle.*
• What shape is part C? (Signal.) *(A) triangle.*
h. Write the letters for each part of the large shape. Don't get fooled.
(Observe children and give feedback.)
i. Check your work.
(Display:) [87:5C]

• (Point to ⬛.) What letter did you write for this part? (Signal.) *B.*

• (Point to ⬛.) What letter did you write for this part? (Signal.) *A.*

• (Point to ⬛.) What letter did you write for this part? (Signal.) *C.*
(Add to show:) [87:5D]

Here's what you should have.

from Lesson 87, Exercise 5

Teaching Note: If students have trouble identifying the letter for a triangle, touch the top triangle in the complex figure and ask, "Is this the bigger triangle or the smaller triangle? So what's the letter of this triangle?" Touch the bottom triangle and ask, "Is this the bigger triangle or the smaller triangle?" "So what's the letter of this triangle?"

On Lesson 95, students label faces of three-dimensional objects with the same letters they used to identify component parts of shapes.

Here's part of the exercise from Lesson 97:

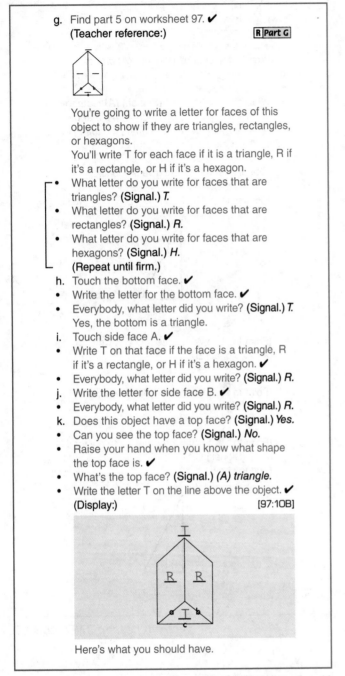

from Lesson 97, Exercise 10

Students identify the whole shape and the parts labeled A, B, C, D. Then they complete equations for a number of different component shapes. In later exercises students work from component parts of a larger figure. They label the parts in the whole figure.

Here is the exercise from Lesson 86:

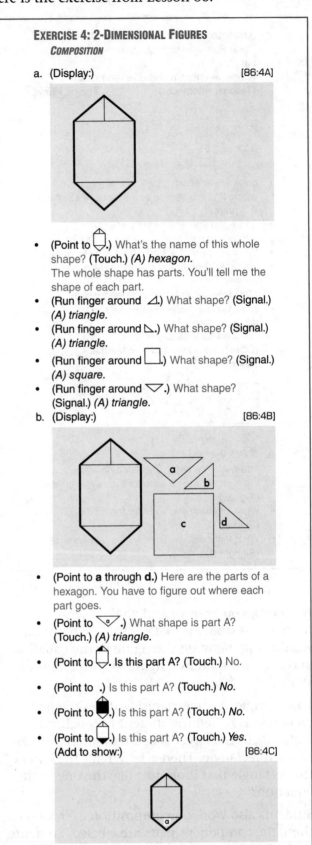

c. (Point to ⬔.) Look at parts B, C, and D.
- Raise your hand when you can tell me the letter of this part. ✔
- Everybody, what's the letter of this part? (Signal.) *D.*
(Add to show:)   [86:4D]

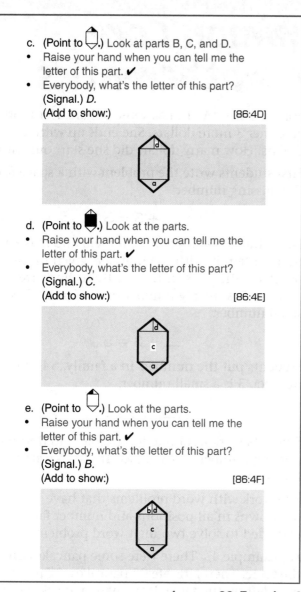

d. (Point to ⬛.) Look at the parts.
- Raise your hand when you can tell me the letter of this part. ✔
- Everybody, what's the letter of this part? (Signal.) *C.*
(Add to show:)   [86:4E]

e. (Point to ⬕.) Look at the parts.
- Raise your hand when you can tell me the letter of this part. ✔
- Everybody, what's the letter of this part? (Signal.) *B.*
(Add to show:)   [86:4F]

Lesson 86, Exercise 4

CMC Level B and Common Core State Standards for Mathematics

According to the Common Core State Standards for Mathematics:

> In Grade 1, instructional time should focus on four critical areas: (1) developing understanding of addition, subtraction, and strategies for addition and subtraction within 20; (2) developing understanding of whole number relationships and place value, including grouping in tens and ones; (3) developing understanding of linear measurement and measuring lengths as iterating length units; and (4) reasoning about attributes of, and composing and decomposing geometric shapes.

CMC Level B meets all the Grade 1 standards. Below is a summary. Note that most of the Common Core State Standards are covered in the major tracks discussed on pages 32–93. Parts of the standards that have not already been addressed are discussed in this section. A comprehensive listing of the standards and where they are met in the program appears on pages 276–282 of Presentation Book 1, pages 296–302 of Presentation Book 2, and pages 388–396 of Presentation Book 3.

Operations and Algebraic Thinking

Common Core State Standards

1.OA1 and 1.OA2	Represent and solve problems involving addition and subtraction.

The problems tell about situations in which values are combined or increased with the unknowns in different positions:

$$\underline{\quad} + 3 = 5$$

$$6 + \underline{\quad} = 9$$

$$2 + 5 = \underline{\quad}$$

For instance, "A girl has some dollars. Her mother gives her 3 more dollars. She ends up with five dollars. How many dollars did she start out with?"

First students write the problem with a space for the missing number:

$$\underline{\quad} + 3 = 5$$

Next students analyze the problem and make a number family with the numbers given in the problem. She ends with more than she started with, so 5 is the big number in the family. 3 is a small number.

$$\underline{\quad} + 3 = 5$$

Students put the numbers in a family. 5 is the big number. 3 is a small number.

$$\xrightarrow{\underline{\qquad}\;3} 5$$

Students work a subtraction problem to find the missing number: 5 – 3 =. (See Tracks, **Number Families.**)

The work with word problems that have unknowns in all positions and number families is extended to solve two-digit word problems.

For example 4, "There were some pancakes on the table. Then girls ate 22 pancakes. 27 pancakes were left on the table. How many pancakes were on the table to start with?"

Students write the symbols for the problem:

$$\underline{\quad} - 22 = 27$$

Students put the numbers in a number family:

$$\xrightarrow{\quad 22 \quad 27 \quad}$$

They work the addition problem:

$$\begin{array}{r} 2\,2 \\ +\ 2\,7 \\ \hline \end{array}$$

They conclude that there were 49 pancakes on the table to start with.

Note that the problems students work involve numbers through 100.

This work is partly achieved by number families and their properties. The families show commutative and inverse properties of addition. The small numbers are added in either order to obtain the big number. The big number is paired with a small number to create subtraction. The basic rules are that to find a missing big number, you add the small numbers. To find a missing small number, you start with the big number and subtract the small number that is shown.

Students work with a variety of problem types that involve equations and statements about more and less. To determine the unknown whole number involving addition or subtraction, students use the signs <, =, >. For example, one side of a statement has 40 + 3 + 10; the other side has 42 + 10. Students determine that there are 43 + 10 on the side with 40 + 3 + 10. This is more than 42 + 10. Students express the relationship between the sides in two ways: with = or ≠ and with <, >, or =.

40 + 3 + 10 ≠ 42 + 10

40 + 3 + 10 > 42 + 10

Number and Operations in Base Ten

Students start at various numbers that are less than 120 and count either forward or backward. Students also write numbers through 1000.

Students write numbers for objects and work problems by counting markers on number lines. For example, students figure out the total number of inches for this "ruler":

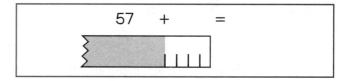

The shaded part is 57 inches. To find the total length, students start at 57 and count for each mark in the unshaded part: *58, 59, 60, 61*. Students conclude that the total length is 61 inches.

Students count the unshaded segments to complete the equation for the "ruler."

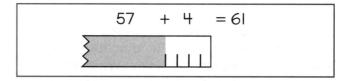

Students also figure out the length of the shaded part of the number line below by counting backward from the end number (65) to the shaded part.

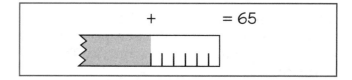

Students learn that 3-digit numerals have a hundreds digit, tens digit, and ones digit. The hundreds digit tells the number of hundreds. The tens digit tells the number of tens. Students learn to represent place value as addition:

400 + 30 + 5 = 435; 50 + 2 = 52. (See Tracks, **Place Value.**)

In the Practice Software, students compare two two-digit numbers on the basis of tens and ones digits.

Common Core State Standards

1.NBT4, 1.NBT5, and 1.NBT6   Use place value understanding and properties of operations to add and subtract.

Place-value addition becomes important when students learn to work problems that rename or carry. If the answer to a problem in the ones column is 12, students identify that 12 is 10 plus 2. The ten belongs in the tens column and is written with the digit 1.

$$
\begin{array}{r}
\overset{1}{3\,9} \\
+\,2\,3 \\
\hline
2
\end{array}
$$

Students also learn several extensions that are based on basic place-value analysis. Given any number less than 990, students are able to identify the number that is ten more or ten less. The reasoning is that the tens digit changes by one when one ten is added or subtracted; however, the ones digit does not change. So, one more ten than 58 is 68. One less ten than 36 is 26.

After students find 10 more or 10 less than a 2-digit number, they are given a chance to explain the reasoning of their computation. (See Tracks, **Counting.**)

The relationship between ones and tens is taught by using a number line that shows numbers 1 through 10:

   1 2 3 4 5 6 7 8 9 10

Adding a zero to each number changes them to corresponding tens numbers:

   10 20 30 40 50 60 70 80 90 100

Students learn to add and subtract multiples of tens as an extension of place-value knowledge.

   For addition: If 6 + 3 = 9, 60 + 30 = 90 (and 600 + 300 = 900)

   For subtraction: if 6 – 2 = 4, 60 – 20 = 40 (and 600 – 200 = 400)

Measurement and Data

Common Core State Standards

1.MD1 and 1.MD2   Measure length indirectly and by iterating length units.

Students are shown that two comparisons imply a third comparison. For example, a picture shows that Jenny is shorter than Lana.

A second picture shows Juan is shorter than Jenny. The question that follows asks about the unobserved comparison: Who is shorter, Juan or Lana? The reasoning is that Juan is shorter than Jenny, Jenny is shorter than Lana, so Juan must be shorter than Lana.

A final picture shows all three people and confirms that Juan is shorter than Lana.

The practice software that accompanies *CMC Level B* permits students to create iterations of measurement units. For instance, the program directs students to measure the length of a pictured object with specified units. Students first identify the correct unit and then place the unit on a template below the picture. They repeat the process until the length of the lined-up units equals the length of the object. Then students type in the number to complete the length statement.

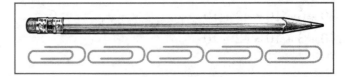

The picture shows that the pencil is 5 paper clips long.

Common Core State Standards

1.MD3   Tell and write time.

Students learn to tell time using both digital and analog displays.

They identify the hour and whether it is "o'clock" or 30 minutes after the hour. They write times using the standard conventions:

For five o'clock, they write: 5:00. For five-thirty, they write: 5:30. The analysis they use assures that they understand that 5:30 is 30 minutes later than 5:00. (See Tracks, **Time.**)

| 1.MD4 | Represent and interpret data. |

Students represent and interpret data by creating frequency charts that show the numbers for different categories of related objects. For instance, an exercise presents a picture of vehicles of different colors (red, blue, green, and yellow). Students count the vehicles of each color and enter the results on a frequency chart. Students write initials for the color names:

	1	2	3	4	5	6
color	G	B		R	Y	

For this example, there are 1 green vehicle, 2 blue vehicles, 4 red vehicles, 5 yellow vehicles. The frequency chart shows that there are no colors with 3 or 6 vehicles.

After completing the chart, students answer questions about the results:

- What's the number of yellow vehicles?
- Are there more yellow vehicles or green vehicles?
- How many more yellow vehicles than red vehicles are there?

The program teaches the steps that students take to answer the various questions.

Geometry

Common Core State Standards

| 1.G1, 1.G2, and 1.G3 | Reason with shapes and their attributes. |

Students learn to identify two-dimensional and three-dimensional shapes. The two-dimensional shapes are rectangles, squares, triangles, pentagons, hexagons, and circles. The three-dimensional shapes are cubes, cones, cylinders, pyramids, and prisms.

Students learn the defining characteristics of the various shapes. For instance, they learn that a cone has a circle at one end and a point at the other. They use these features to distinguish shapes. For instance, a cylinder is presented and students identify the shape and then answer questions that are related to the defining features of a cylinder. "How do you know this isn't a cone?" Students express the idea that it doesn't have a point at one end and a circle at the other. So it can't be a cone. In the same way, they defend their judgment that the object is a cylinder by answering the question: "How do you know this is a cylinder?"

They express the idea that it is not a flat object, and it has a point at one end and a circle at the other.

Students learn that prisms share properties with cylinders. Both ends are the same size. Prisms, however, don't have circles at each end, but familiar shapes with straight sides—triangles, rectangles, pentagons and hexagons. For prisms, rectangles connect the congruent shapes at both ends.

Students learn that pyramids share properties with cones. Both have a point at one end. At the other end however, pyramids have straight-sided shapes—triangle, rectangle, pentagon, etc.—instead of a circle. For pyramids, triangles connect with the straight-sided shape. Again, students use the properties of the pyramid shown to answer questions like, "How do you know this shape is not a prism?" (It doesn't have the same shape on both ends. It has a point at one end. The sides are triangles not rectangles.)

Students use the software program that accompanies *CMC Level B* to compose the various three-dimensional shapes using illustrated components for the ends and the middle. To create a pyramid with the rectangular base, students would select the square for the base and then move four triangles to form the sides.

The software program also has exercises that allow students to distinguish between defining and non-defining attributes of shapes (i.e., color versus number of sides) and to partition circles and rectangles into segments that are the same size. Students create circles and squares that are divided into fourths and halves. Students' work with fractions teaches them about halves, half, quarters, and fourths. (See Tracks, **Fractions.**)

Appendix A

Placement Test

Placement Test

NOTE: There are two sections to this test. Section I is on the first page of the test sheet. Present Section I to all students who are being considered for placement in *CMC Level B*. After presenting Section I, collect and grade Section I of the test sheet. Use the placement criteria to determine the next assessment that should be presented to each student.

Distribute the second page of the test sheet to students who meet the criterion for taking Section II of the Placement Test. Present Section II. After presenting Section II, collect and grade Section II. Use the placement criteria to determine the placement or next assessment for each of the students who took Section II.

- For this test, each student will need a pencil.
- Try to arrange students so they cannot look at other students' responses.
- Make sure student's name is on their test sheet.
- After students have completed the test, collect the test sheets and mark students' responses.
- Allow no more than the time indicated for students to complete each part. Students who can successfully work the problems but require more than the time allotted by the test will not be successful in *CMC Level B*.
- After grading students' test sheets, record their performance on the Placement Test Summary Sheet. Use the placement criteria to determine each student's placement or the assessment that should be administered next.

TEACHER PRESENTATION

SECTION I

Part 1 (5 points possible)

a. Touch the diamond on your test sheet. ✔

I'm going to say numbers. You'll say the number, then you'll write it.
- Touch the first box next to the diamond. ✔
- That's where you'll write the first number.
- Listen: 7. What number did I say? (Signal.) *7.*
- Write 7 in the first box.
  (Observe but do not give feedback.)

b. Touch the next box. ✔
- The next number is 3. What number? (Signal.) *3.*
- Write 3.
  (Observe but do not give feedback.)

c. Touch the next box. ✔
- The next number is 9. What number? (Signal.) *9.*
- Write 9.
  (Observe but do not give feedback.)

d. Touch the next box. ✔
- The next number is 5. What number? (Signal.) *5.*
- Write 5.
  (Observe but do not give feedback.)

e. Touch the last box. ✔
- The last number for this part is zero. What number? (Signal.) *zero.*
- Write zero.
  (Observe but do not give feedback.)

Part 2 (8 points possible)

a. Touch the moon on your test sheet. ✔

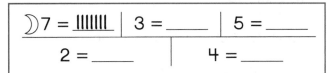

$$\text{☽}7 = \underline{|||||||} \quad | \quad 3 = \underline{} \quad | \quad 5 = \underline{}$$
$$2 = \underline{} \quad | \quad 4 = \underline{}$$

Each problem shows a number and an equal sign. You're going to make lines to make the sides equal. The first problem is already worked. It shows 7 equals, so there are 7 lines on the other side.
- Touch the equals next to 3. ✔
  How many lines will you make to complete that equation? (Signal.) *3.*
- Touch the equals next to 5. ✔
  How many lines will you make to complete that equation? (Signal.) *5.*
- Touch the equals next to 2. ✔
  How many lines will you make to complete that equation? (Signal.) *2.*
- Touch the equals next to 4. ✔
  How many lines will you make to complete that equation? (Signal.) *4.*
  (Repeat tasks that were not firm.)

b. Make lines to complete each equation in this part.
  (Observe but do not give feedback.)
  (After students are finished or after no more than 3 minutes, present Part 3.)

Part 3 (6 points possible)

a. Touch the problem 8 plus 1. ✔

$8 + 1 =$	$4 + 1 =$	6	9
$5 + 0 =$	$10 + 0 =$	$\underline{+1}$	$\underline{+0}$

Some of the problems in this part are written in rows. Some of them are written in columns. Each of the problems in this part is either plus 1 or plus zero. I'll read the problems written in rows first. You touch them as I read them.

- The first problem is 8 plus 1. ✔
- The next problem is 4 plus 1. ✔
- The first problem in the next row is 5 plus 0. ✔
- The next problem is 10 plus 0. ✔
- The first column problem for this part is 6 plus 1. ✔
- The next column problem is 9 + 0. ✔

b. Work the plus-1 problems and the plus-zero problems in this part.
  (Observe students but do not give feedback.)
  (After students are finished or after no more than 4 minutes, present Part 4.)

Part 4 (6 points possible)

a. Touch the problem 10 minus 1. Another way to read that problem is 10 take away 1. Touch it. ✔

$10 - 1 =$	$5 - 1 =$	8	6
$2 - 1 =$	$3 - 0 =$	$\underline{-1}$	$\underline{-0}$

Some of the problems in this part are written in rows. Some of them are written in columns. Each of the problems in this part is either minus 1 or minus 0. I'll read the problems written in rows first. You touch them as I read them.

- The first problem is 10 minus 1. ✔
- The next problem is 5 minus 1. ✔
- The first problem in the next row is 2 minus 1. ✔
- The next problem is 3 minus zero. ✔
- The first column problem for this part is 8 minus 1. ✔
- The next column problem is 6 minus 0. ✔

b. Work the minus-1 problems and the minus-zero problems in this part.
  (Observe students but do not give feedback.)
  (After students are finished or after no more than 4 minutes, collect test sheets, and grade them. For students who meet the criterion for taking Section II, distribute page 2 of test sheets and present Section II.)

SECTION II

Part 5 (5 points possible)

a. Touch the star on your test sheet. ✔

I'm going to say numbers. You'll say each number, then you'll write it.

- Touch the first box next to the star. ✔
  That's where you'll write the first number.
- Listen: 47. What number did I say? (Signal.) *47.*
- Write 47 in the first box.
  (Observe but do not give feedback.)

b. Touch the next box. ✔

- The next number is 64. What number? (Signal.) *64.*
- Write 64.
  (Observe but do not give feedback.)

c. Touch the next box. ✔

- The next number is 80. What number? (Signal.) *80.*
- Write 80.
  (Observe but do not give feedback.)

d. Touch the next box. ✔

- The next number is 18. What number? (Signal.) *18.*
- Write 18.
  (Observe but do not give feedback.)

e. Touch the last box. ✔

- The last number is 13. What number? (Signal.) *13.*
- Write 13.
  (Observe but do not give feedback.)

Part 6 (4 points possible)

a. Touch the heart on your test sheet. ✔

I'm going to say equations. You'll say each equation. Then you'll write it.

- Touch the line next to the heart. ✔
  That's where you'll write the first equation.
- Listen to the first equation: 5 plus 2 equals 7. Say that equation. (Signal.) *5 plus 2 equals 7.*
  (Repeat until firm.)
- Write 5 plus 2 equals 7 on the first line next to the heart.
  (Observe but do not give feedback.)

b. Touch the next line. ✔

- The equation you'll write for that line is 9 minus 3 equals 6. Say that equation. (Signal.) *9 minus 3 equals 6.*
  (Repeat until firm.)
- Another way of saying 9 minus 3 equals 6 is 9 take away 3 equals 6. Say 9 take away 3 equals 6. (Signal.) *9 take away 3 equals 6.*
  (Repeat until firm.)
- Write 9 minus 3 equals 6.
  (Observe but do not give feedback.)

Part 7 (6 points possible)

a. Touch the problem 4 plus 2. ✔

$4 + 2 =$	$9 + 2 =$	$\begin{array}{r} 15 \\ +\ 1 \\ \hline \end{array}$	$\begin{array}{r} 39 \\ +\ 0 \\ \hline \end{array}$
$27 + 1 =$	$13 + 2 =$		

Some of the problems in this part are written in rows. Some of them are written in columns. All of the problems in this part plus-2, plus-1, or plus zero. Work the plus problems in this part.
(Observe but do not give feedback.)
(After students are finished or after no more than 4 minutes, collect test sheets, and grade them.)

Part 8 (12 points possible)

a. Touch the problem 20 plus 10. ✔

$20 + 10 =$	$60 + 3 =$	$\begin{array}{r} 8 \\ +40 \\ \hline \end{array}$	$\begin{array}{r} 2 \\ +58 \\ \hline \end{array}$
$27 + 10 =$	$20 + 5 =$		

Work the problems in this part.
(Observe but do not give feedback.)
(After students are finished or after no more than 4 minutes, collect test sheets and grade them. Use the placement criteria to determine student placement or additional assessments.)

SCORING NOTES

When grading the Placement Test, accept reversed digits. $\varepsilon = 3$.
Do not accept transposed digits. $12 \neq 21$.
Do not accept transposed symbols. $5 + 2 \neq 5\ 2\ +$.

OVERVIEW

SECTION	SCORE	ACTION
I	0–15	Test for placement in an entry level or K program (*CMC Level A, Distar Arithmetic*)
	16–25	Present Section II
II	0–19	Place on Lesson 1 of *CMC Level B*
	20–27	Place on Lesson 16 of *CMC Level B*
	(26–27)	(Assess more advanced placement)

SECTION I, PARTS 1–4

Part 1: Students earn 1 point for writing the correct number in each box in Part 1. Students earn zero points for a box that does not have the correct number in it or for a box with more than the correct number in it.

Students can earn 5 points for Part 1.

Part 2: Students earn 2 points for completing each equation with the correct number of lines. Students earn zero points for equations that do not have the correct number of lines.

Students can earn 8 points for Part 2.

Part 3: Students earn 1 point for each correct answer. Students earn zero points for equations that do not have only the correct answer written.

Students can earn 6 points for Part 3.

Part 4: Students earn 1 point for each correct answer. Students earn zero points for equations that do not have only the correct answer written.

Students can earn 6 points for Part 4.

Section I Criterion

The total number of points possible for Section I is 25. For students who score 15 or fewer points in Section I, test them for placement in a Kindergarten math sequence. For students who score above 15 points, present Section II.

Section II, Parts 5–8

Part 5: Students earn 1 point for writing the correct number in each box for part 5. Students earn zero points for boxes that do not have the correct number in it or for boxes with symbols other than the correct number in it.

Students can earn 5 points for Part 5.

Part 6: Students earn 2 points for writing the correct equation on each line. Students earn zero points for an equation if it doesn't have only the correct digits in the correct order.

Students can earn 0 points, 2 points, or 4 points for Part 6.

Part 7: Students earn 1 point for each correct answer. Students earn zero points for incorrect answers or answers with symbols other than the correct digits.

Students can earn 6 points for Part 7.

Part 8: Students earn 2 points for each correct answer. Students earn zero points for incorrect answers or answers with symbols other than the correct digits.

Students can earn 12 points for Part 8.

Section II Criterion

The total number of points possible for Section II is 27. For students who score 19 or fewer points in Section II, begin instruction on Lesson 1 of the Pre-program. For students who score 20 or more points, they should begin instruction on Lesson 16 of the program. Consider assessing children who score 26 or 27 for placement in the middle of *CMC Level B* sequence or in a second-grade program.

Placement Test
Section I

□ □ □ □ □

)7 = ‖‖‖‖‖	3 = _____	5 = _____

2 = _____   4 = _____

8 + 1 =	4 + 1 =	6 + 1 ___	9 + 0 ___
5 + 0 =	10 + 0 =		
10 − 1 =	5 − 1 =	8 − 1 ___	6 − 0 ___
2 − 1 =	3 − 0 =		

Placement Test
Section II

4 + 2 =	9 + 2 =	15 + 1	39 + 0
27 + 1 =	13 + 2 =		
20 + 10 =	60 + 3 =	8 +40	2 +58
27 + 10 =	20 + 5 =		

Connecting Math Concepts

CMC Level B Placement Test Answer Key, Section I

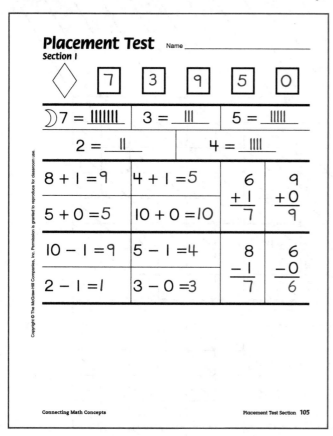

CMC Level B Placement Test Answer Key, Section II

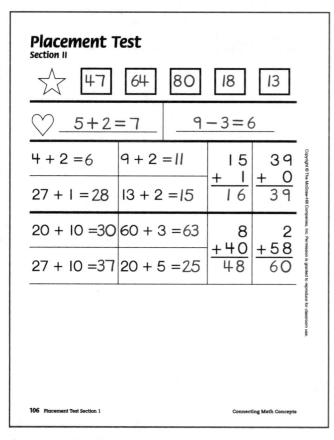

Placement Test
Summary Sheet

Name

Section I																
Total points for **Part 1**	5	5	5	5	5	5	5	5	5	5	5	5	5	5	5	5
Total points for **Part 2**	8	8	8	8	8	8	8	8	8	8	8	8	8	8	8	8
Total points for **Part 3**	6	6	6	6	6	6	6	6	6	6	6	6	6	6	6	6
Total points for **Part 4**	6	6	6	6	6	6	6	6	6	6	6	6	6	6	6	6
Total score for **Section I**	25	25	25	25	25	25	25	25	25	25	25	25	25	25	25	25
Test for entry level program	Y/N	Y/N	Y/N	Y/N	Y/N	Y/N	Y/N	Y/N	Y/N	Y/N	Y/N	Y/N	Y/N	Y/N	Y/N	Y/N
Present Section II	Y/N	Y/N	Y/N	Y/N	Y/N	Y/N	Y/N	Y/N	Y/N	Y/N	Y/N	Y/N	Y/N	Y/N	Y/N	Y/N
Section II																
Total points for **Part 5**	5	5	5	5	5	5	5	5	5	5	5	5	5	5	5	5
Total points for **Part 6**	4	4	4	4	4	4	4	4	4	4	4	4	4	4	4	4
Total points for **Part 7**	6	6	6	6	6	6	6	6	6	6	6	6	6	6	6	6
Total points for **Part 8**	12	12	12	12	12	12	12	12	12	12	12	12	12	12	12	12
Total score for **Section II**	27	27	27	27	27	27	27	27	27	27	27	27	27	27	27	27
Start on **CMC Level B L1**	Y/N	Y/N	Y/N	Y/N	Y/N	Y/N	Y/N	Y/N	Y/N	Y/N	Y/N	Y/N	Y/N	Y/N	Y/N	Y/N
Start on **CMC Level B L16**	Y/N	Y/N	Y/N	Y/N	Y/N	Y/N	Y/N	Y/N	Y/N	Y/N	Y/N	Y/N	Y/N	Y/N	Y/N	Y/N
Assess more advanced placement	Y/N	Y/N	Y/N	Y/N	Y/N	Y/N	Y/N	Y/N	Y/N	Y/N	Y/N	Y/N	Y/N	Y/N	Y/N	Y/N

Connecting Math Concepts

Appendix B

Reproducible Cumulative Test Summary Sheets

	Name															
Total points for **Part 1**	10	10	10	10	10	10	10	10	10	10	10	10	10	10	10	10
Pass child if score is at least 9/10.	P N	P N	P N	P N	P N	P N	P N	P N	P N	P N	P N	P N	P N	P N	P N	P N
Total points for **Part 2**	25	25	25	25	25	25	25	25	25	25	25	25	25	25	25	25
Pass child if score is at least 22/25.	P N	P N	P N	P N	P N	P N	P N	P N	P N	P N	P N	P N	P N	P N	P N	P N
Total points for **Part 3**	6	6	6	6	6	6	6	6	6	6	6	6	6	6	6	6
Pass child if score is at least 5/6.	P N	P N	P N	P N	P N	P N	P N	P N	P N	P N	P N	P N	P N	P N	P N	P N
Total points for **Part 4**	8	8	8	8	8	8	8	8	8	8	8	8	8	8	8	8
Pass child if score is at least 7/8.	P N	P N	P N	P N	P N	P N	P N	P N	P N	P N	P N	P N	P N	P N	P N	P N
Total points for **Part 5**	6	6	6	6	6	6	6	6	6	6	6	6	6	6	6	6
Pass child if score is at least 5/6.	P N	P N	P N	P N	P N	P N	P N	P N	P N	P N	P N	P N	P N	P N	P N	P N
Total points for **Part 6**	8	8	8	8	8	8	8	8	8	8	8	8	8	8	8	8
Pass child if score is at least 7/8.	P N	P N	P N	P N	P N	P N	P N	P N	P N	P N	P N	P N	P N	P N	P N	P N
Total points for **Part 7**	3	3	3	3	3	3	3	3	3	3	3	3	3	3	3	3
Pass child if score is 3/3.	P N	P N	P N	P N	P N	P N	P N	P N	P N	P N	P N	P N	P N	P N	P N	P N
Total points for **Part 8**	8	8	8	8	8	8	8	8	8	8	8	8	8	8	8	8
Pass child if score is at least 7/8.	P N	P N	P N	P N	P N	P N	P N	P N	P N	P N	P N	P N	P N	P N	P N	P N
Total points for **Part 9**	10	10	10	10	10	10	10	10	10	10	10	10	10	10	10	10
Pass child if score is at least 8/10.	P N	P N	P N	P N	P N	P N	P N	P N	P N	P N	P N	P N	P N	P N	P N	P N
Total points for **Part 10**	5	5	5	5	5	5	5	5	5	5	5	5	5	5	5	5
Pass child if score is 5/5.	P N	P N	P N	P N	P N	P N	P N	P N	P N	P N	P N	P N	P N	P N	P N	P N
Total points for **Part 11**	4	4	4	4	4	4	4	4	4	4	4	4	4	4	4	4
Pass child if score is 4/4.	P N	P N	P N	P N	P N	P N	P N	P N	P N	P N	P N	P N	P N	P N	P N	P N
Total points for **Part 12**	25	25	25	25	25	25	25	25	25	25	25	25	25	25	25	25
Pass child if score is at least 22/25.	P N	P N	P N	P N	P N	P N	P N	P N	P N	P N	P N	P N	P N	P N	P N	P N
Total points for **Part 13**	15	15	15	15	15	15	15	15	15	15	15	15	15	15	15	15
Pass child if score is at least 14/15.	P N	P N	P N	P N	P N	P N	P N	P N	P N	P N	P N	P N	P N	P N	P N	P N
Total points for **Part 14**	4	4	4	4	4	4	4	4	4	4	4	4	4	4	4	4
Pass child if score is 4/4.	P N	P N	P N	P N	P N	P N	P N	P N	P N	P N	P N	P N	P N	P N	P N	P N
Total points for **Part 15**	8	8	8	8	8	8	8	8	8	8	8	8	8	8	8	8
Pass child if score is 8/8.	P N	P N	P N	P N	P N	P N	P N	P N	P N	P N	P N	P N	P N	P N	P N	P N

	Name															
Total points for **Part 16**	4	4	4	4	4	4	4	4	4	4	4	4	4	4	4	4
Pass child if score is 4/4.	P N	P N	P N	P N	P N	P N	P N	P N	P N	P N	P N	P N	P N	P N	P N	P N
Total points for **Part 17**	10	10	10	10	10	10	10	10	10	10	10	10	10	10	10	10
Pass child if score is at least 8/10.	P N	P N	P N	P N	P N	P N	P N	P N	P N	P N	P N	P N	P N	P N	P N	P N
Total number of points	159	159	159	159	159	159	159	159	159	159	159	159	159	159	159	159

	Name															
Total points for **Part 1**	8	8	8	8	8	8	8	8	8	8	8	8	8	8	8	8
Pass child if score is at least 7/8.	P N	P N	P N	P N	P N	P N	P N	P N	P N	P N	P N	P N	P N	P N	P N	P N
Total points for **Part 2**	8	8	8	8	8	8	8	8	8	8	8	8	8	8	8	8
Pass child if score is at least 7/8.	P N	P N	P N	P N	P N	P N	P N	P N	P N	P N	P N	P N	P N	P N	P N	P N
Total points for **Part 3**	3	3	3	3	3	3	3	3	3	3	3	3	3	3	3	3
Pass child if score is 3/3.	P N	P N	P N	P N	P N	P N	P N	P N	P N	P N	P N	P N	P N	P N	P N	P N
Total points for **Part 4**	8	8	8	8	8	8	8	8	8	8	8	8	8	8	8	8
Pass child if score is at least 6/8.	P N	P N	P N	P N	P N	P N	P N	P N	P N	P N	P N	P N	P N	P N	P N	P N
Total points for **Part 5**	30	30	30	30	30	30	30	30	30	30	30	30	30	30	30	30
Pass child if score is at least 25/30.	P N	P N	P N	P N	P N	P N	P N	P N	P N	P N	P N	P N	P N	P N	P N	P N
Section A Subtotal	57	57	57	57	57	57	57	57	57	57	57	57	57	57	57	57
Section A Parts Passed	5	5	5	5	5	5	5	5	5	5	5	5	5	5	5	5
Present Section B if parts passed is at least 4/5.	Y N	Y N	Y N	Y N	Y N	Y N	Y N	Y N	Y N	Y N	Y N	Y N	Y N	Y N	Y N	Y N
Total points for **Part 6**	4	4	4	4	4	4	4	4	4	4	4	4	4	4	4	4
Pass child if score is 4/4.	P N	P N	P N	P N	P N	P N	P N	P N	P N	P N	P N	P N	P N	P N	P N	P N
Total points for **Part 7**	9	9	9	9	9	9	9	9	9	9	9	9	9	9	9	9
Pass child if score is 9/9.	P N	P N	P N	P N	P N	P N	P N	P N	P N	P N	P N	P N	P N	P N	P N	P N
Total points for **Part 8**	8	8	8	8	8	8	8	8	8	8	8	8	8	8	8	8
Pass child if score is at least 7/8.	P N	P N	P N	P N	P N	P N	P N	P N	P N	P N	P N	P N	P N	P N	P N	P N
Total points for **Part 9**	11	11	11	11	11	11	11	11	11	11	11	11	11	11	11	11
Pass child if score is 11/11.	P N	P N	P N	P N	P N	P N	P N	P N	P N	P N	P N	P N	P N	P N	P N	P N
Total points for **Part 10**	7	7	7	7	7	7	7	7	7	7	7	7	7	7	7	7
Pass child if score is 7/7.	P N	P N	P N	P N	P N	P N	P N	P N	P N	P N	P N	P N	P N	P N	P N	P N
Total points for **Part 11**	6	6	6	6	6	6	6	6	6	6	6	6	6	6	6	6
Pass child if score is at least 5/6.	P N	P N	P N	P N	P N	P N	P N	P N	P N	P N	P N	P N	P N	P N	P N	P N
Total points for **Part 12**	10	10	10	10	10	10	10	10	10	10	10	10	10	10	10	10
Pass child if score is at least 8/10.	P N	P N	P N	P N	P N	P N	P N	P N	P N	P N	P N	P N	P N	P N	P N	P N
Total points for **Part 13**	4	4	4	4	4	4	4	4	4	4	4	4	4	4	4	4
Pass child if score is 4/4.	P N	P N	P N	P N	P N	P N	P N	P N	P N	P N	P N	P N	P N	P N	P N	P N
Total points for **Part 14**	10	10	10	10	10	10	10	10	10	10	10	10	10	10	10	10
Pass child if score is at least 7/10.	P N	P N	P N	P N	P N	P N	P N	P N	P N	P N	P N	P N	P N	P N	P N	P N

Name →															
Total points for Part 15	4	4	4	4	4	4	4	4	4	4	4	4	4	4	4
Pass child if score is 4/4.	P/N	P/N	P/N	P/N	P/N	P/N	P/N	P/N	P/N	P/N	P/N	P/N	P/N	P/N	P/N
Total points for Part 16	3	3	3	3	3	3	3	3	3	3	3	3	3	3	3
Pass child if score is at least 2/3.	P/N	P/N	P/N	P/N	P/N	P/N	P/N	P/N	P/N	P/N	P/N	P/N	P/N	P/N	P/N
Total points for Part 17	10	10	10	10	10	10	10	10	10	10	10	10	10	10	10
Pass child if score is at least 8/10.	P/N	P/N	P/N	P/N	P/N	P/N	P/N	P/N	P/N	P/N	P/N	P/N	P/N	P/N	P/N
Total points for Part 18	6	6	6	6	6	6	6	6	6	6	6	6	6	6	6
Pass child if score is at least 5/6.	P/N	P/N	P/N	P/N	P/N	P/N	P/N	P/N	P/N	P/N	P/N	P/N	P/N	P/N	P/N
Total points for Part 19	6	6	6	6	6	6	6	6	6	6	6	6	6	6	6
Pass child if score is at least 5/6.	P/N	P/N	P/N	P/N	P/N	P/N	P/N	P/N	P/N	P/N	P/N	P/N	P/N	P/N	P/N
Total points for Part 20	9	9	9	9	9	9	9	9	9	9	9	9	9	9	9
Pass child if score is at least 7/9.	P/N	P/N	P/N	P/N	P/N	P/N	P/N	P/N	P/N	P/N	P/N	P/N	P/N	P/N	P/N
Total points for Part 21	4	4	4	4	4	4	4	4	4	4	4	4	4	4	4
Pass child if score is 4/4.	P/N	P/N	P/N	P/N	P/N	P/N	P/N	P/N	P/N	P/N	P/N	P/N	P/N	P/N	P/N
Total points for Part 22	6	6	6	6	6	6	6	6	6	6	6	6	6	6	6
Pass child if score is at least 5/6.	P/N	P/N	P/N	P/N	P/N	P/N	P/N	P/N	P/N	P/N	P/N	P/N	P/N	P/N	P/N
Total points for Part 23	6	6	6	6	6	6	6	6	6	6	6	6	6	6	6
Pass child if score is at least 5/6.	P/N	P/N	P/N	P/N	P/N	P/N	P/N	P/N	P/N	P/N	P/N	P/N	P/N	P/N	P/N
Total points for Part 24	20	20	20	20	20	20	20	20	20	20	20	20	20	20	20
Pass child if score is at least 17/20.	P/N	P/N	P/N	P/N	P/N	P/N	P/N	P/N	P/N	P/N	P/N	P/N	P/N	P/N	P/N
Total points for Part 25	15	15	15	15	15	15	15	15	15	15	15	15	15	15	15
Pass child if score is at least 12/15.	P/N	P/N	P/N	P/N	P/N	P/N	P/N	P/N	P/N	P/N	P/N	P/N	P/N	P/N	P/N
Total points for Part 26	4	4	4	4	4	4	4	4	4	4	4	4	4	4	4
Pass child if score is 4/4.	P/N	P/N	P/N	P/N	P/N	P/N	P/N	P/N	P/N	P/N	P/N	P/N	P/N	P/N	P/N
Total points for Part 27	15	15	15	15	15	15	15	15	15	15	15	15	15	15	15
Pass child if score is at least 12/15.	P/N	P/N	P/N	P/N	P/N	P/N	P/N	P/N	P/N	P/N	P/N	P/N	P/N	P/N	P/N
Total points for Part 28	15	15	15	15	15	15	15	15	15	15	15	15	15	15	15
Pass child if score is at least 12/15.	P/N	P/N	P/N	P/N	P/N	P/N	P/N	P/N	P/N	P/N	P/N	P/N	P/N	P/N	P/N
Section B Subtotal	192	192	192	192	192	192	192	192	192	192	192	192	192	192	192
Total number of points	249	249	249	249	249	249	249	249	249	249	249	249	249	249	249

Appendix C

Reproducible Mastery Test Summary Sheets

Mastery Test 1
Individually Administered
Section Summary Sheet

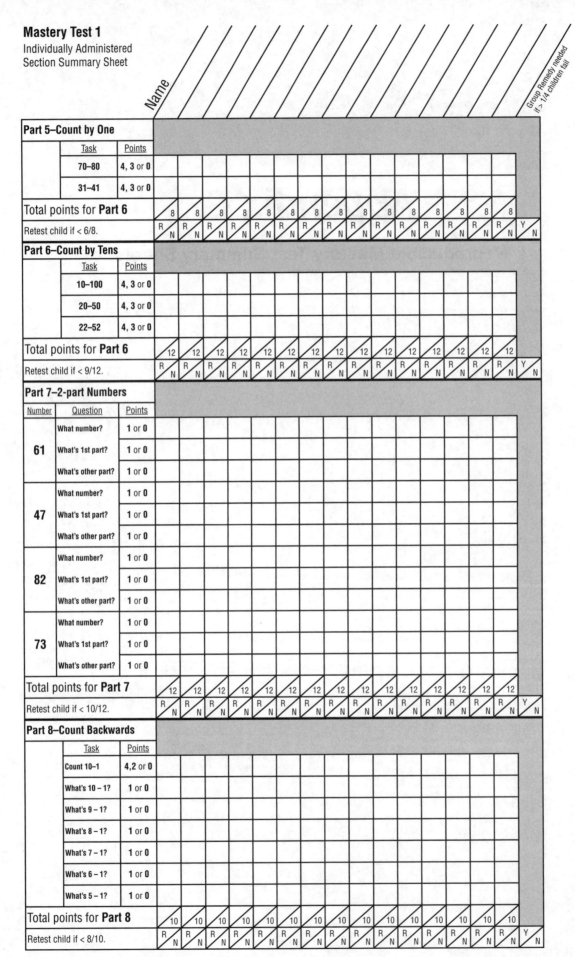

Name

Group Remedy needed if > 1/4 children fail

Part 5–Count by One

Task	Points																
70–80	4, 3 or 0																
31–41	4, 3 or 0																

Total points for **Part 6**	8	8	8	8	8	8	8	8	8	8	8	8	8	8	8	
Retest child if < 6/8.	R N	R N	R N	R N	R N	R N	R N	R N	R N	R N	R N	R N	R N	R N	R N	Y N

Part 6–Count by Tens

Task	Points																
10–100	4, 3 or 0																
20–50	4, 3 or 0																
22–52	4, 3 or 0																

Total points for **Part 6**	12	12	12	12	12	12	12	12	12	12	12	12	12	12	12	
Retest child if < 9/12.	R N	R N	R N	R N	R N	R N	R N	R N	R N	R N	R N	R N	R N	R N	R N	Y N

Part 7–2-part Numbers

Number	Question	Points																
61	What number?	1 or 0																
	What's 1st part?	1 or 0																
	What's other part?	1 or 0																
47	What number?	1 or 0																
	What's 1st part?	1 or 0																
	What's other part?	1 or 0																
82	What number?	1 or 0																
	What's 1st part?	1 or 0																
	What's other part?	1 or 0																
73	What number?	1 or 0																
	What's 1st part?	1 or 0																
	What's other part?	1 or 0																

Total points for **Part 7**	12	12	12	12	12	12	12	12	12	12	12	12	12	12	12	
Retest child if < 10/12.	R N	R N	R N	R N	R N	R N	R N	R N	R N	R N	R N	R N	R N	R N	R N	Y N

Part 8–Count Backwards

Task	Points																
Count 10–1	4, 2 or 0																
What's 10 – 1?	1 or 0																
What's 9 – 1?	1 or 0																
What's 8 – 1?	1 or 0																
What's 7 – 1?	1 or 0																
What's 6 – 1?	1 or 0																
What's 5 – 1?	1 or 0																

Total points for **Part 8**	10	10	10	10	10	10	10	10	10	10	10	10	10	10	10	
Retest child if < 8/10.	R N	R N	R N	R N	R N	R N	R N	R N	R N	R N	R N	R N	R N	R N	R N	Y N

Connecting Math Concepts

Mastery Test 2
Individually Administered
Section Summary Sheet

Name

Group Remedy needed if > 1/4 children fail

Part 8–Count by Tens

	Task	Points														
	50–90	**4, 3 or 0**														
Plus Tens	**53–93**	**4, 3 or 0**														
Plus Tens	**57–97**	**4, 3 or 0**														
Total points for **Part 8**			12	12	12	12	12	12	12	12	12	12	12	12	12	12
Retest child if < 9/12.			R N	R N	R N	R N	R N	R N	R N	R N	R N	R N	R N	R N	R N	Y N

Part 9–Count Backward

	Task	Points														
	8–1	**3, 2 or 0**														
	48–41	**3, 2 or 0**														
	49–40	**3, 2 or 0**														
Total points for **Part 9**			9	9	9	9	9	9	9	9	9	9	9	9	9	9
Retest child if < 6/9.			R N	R N	R N	R N	R N	R N	R N	R N	R N	R N	R N	R N	R N	Y N

Part 10–Reading Numbers

Number	What's __?	Points														
168	underlined part	**1 or 0**														
	whole number	**1 or 0**														
116	underlined part	**1 or 0**														
	whole number	**1 or 0**														
160	underlined part	**1 or 0**														
	whole number	**1 or 0**														
106	underlined part	**1 or 0**														
	whole number	**1 or 0**														
140	underlined part	**1 or 0**														
	whole number	**1 or 0**														
104	underlined part	**1 or 0**														
	whole number	**1 or 0**														
Total points for **Part 10**			12	12	12	12	12	12	12	12	12	12	12	12	12	12
Retest child if < 10/12.			R N	R N	R N	R N	R N	R N	R N	R N	R N	R N	R N	R N	R N	Y N

Part 11–Number Family

Question	Points														
How many numbers?	**1 or 0**														
How many small?	**1 or 0**														
How many big?	**1 or 0**														
5 2→7 small numbers?	**1 or 0**														
big number?	**1 or 0**														
3 1→4 small numbers?	**1 or 0**														
big number?	**1 or 0**														
Total points for **Part 11**		7	7	7	7	7	7	7	7	7	7	7	7	7	7
Retest child if < 7/7.		R N	R N	R N	R N	R N	R N	R N	R N	R N	R N	R N	R N	R N	Y N

Mastery Test 3
Individually Administered
Section Summary Sheet

Name

Individually Administered

Part 9–Counting

	Task	Points																
By hundreds	100–900	**4, 3** or **0**																
Backward	70–60	**4, 3** or **0**																
Backward	40–30	**4, 3** or **0**																
Total points for **Part 9**			12	12	12	12	12	12	12	12	12	12	12	12	12	12	12	
Retest child if < 9/12.			R/N	R/N	R/N	R/N	R/N	R/N	R/N	R/N	R/N	R/N	R/N	R/N	R/N	R/N	R/N	Y/N

Part 10–Digits

Number	Question	Points																
	How many digits?	**1** or **0**																
54	What's 1s digit?	**1** or **0**																
	What's 10s digit?	**1** or **0**																
	How many digits?	**1** or **0**																
375	What's 1s digit?	**1** or **0**																
	What's 10s digit?	**1** or **0**																
	What's 100s digit?	**1** or **0**																
9	How many digits?	**1** or **0**																
	What's 1s digit?	**1** or **0**																
	How many digits?	**1** or **0**																
520	What's 1s digit?	**1** or **0**																
	What's 10s digit?	**1** or **0**																
	What's 100s digit?	**1** or **0**																
Total points for **Part 10**			13	13	13	13	13	13	13	13	13	13	13	13	13	13	13	
Retest child if < 11/13.			R/N	R/N	R/N	R/N	R/N	R/N	R/N	R/N	R/N	R/N	R/N	R/N	R/N	R/N	R/N	Y/N

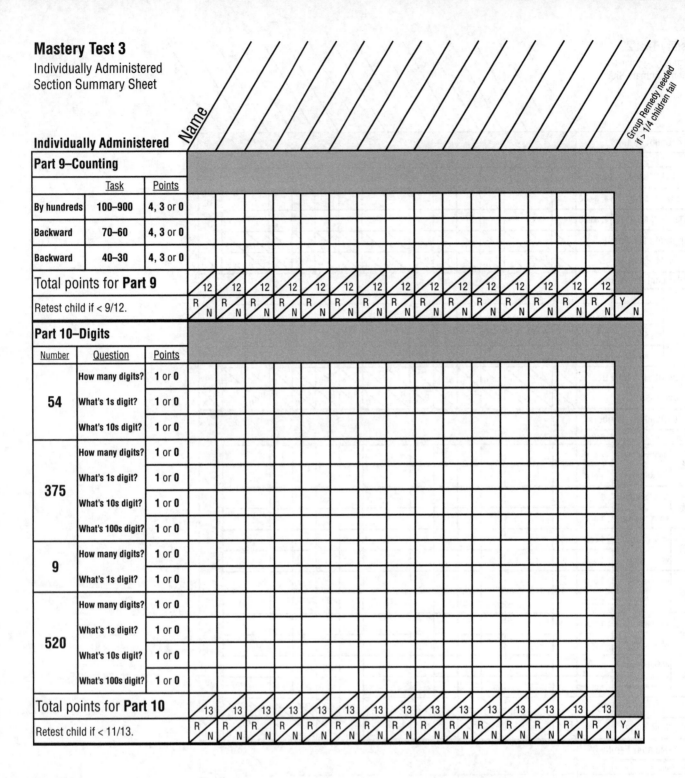

Remedy Summary—Group Summary of Test Performance

Note: Test remedies are specified in the *Answer Key.* All Mastery Tests are 100 points so the student's score is also the total percentage.

Name	Test 1 Check parts not passed				Total %	Test 2 Check parts not passed							Total %	Test 3 Check parts not passed								Total %
	1	2	3	4		1	2	3	4	5	6	7		1	2	3	4	5	6	7	8	
1.																						
2.																						
3.																						
4.																						
5.																						
6.																						
7.																						
8.																						
9.																						
10.																						
11.																						
12.																						
13.																						
14.																						
15.																						
16.																						
17.																						
18.																						
19.																						
20.																						
21.																						
22.																						
23.																						
24.																						
25.																						
26.																						
27.																						
28.																						
29.																						
30.																						

	Test 1	Test 2	Test 3
Number of students Not Passed = NP			
Total number of students = T			
Remedy needed if NP/T = 25% or more			

Remedy Summary—Group Summary of Test Performance

Note: Test remedies are specified in the *Answer Key.* All Mastery Tests are 100 points so the student's score is also the total percentage.

	Test 4						Test 5										Test 6								
Name	Check parts not passed					Total %	Check parts not passed									Total %	Check parts not passed								Total %
	1	2	3	4	5		1	2	3	4	5	6	7	8	9		1	2	3	4	5	6	7		
1.																									
2.																									
3.																									
4.																									
5.																									
6.																									
7.																									
8.																									
9.																									
10.																									
11.																									
12.																									
13.																									
14.																									
15.																									
16.																									
17.																									
18.																									
19.																									
20.																									
21.																									
22.																									
23.																									
24.																									
25.																									
26.																									
27.																									
28.																									
29.																									
30.																									
Number of students Not Passed = NP																									
Total number of students = T																									
Remedy needed if NP/T = 25% or more																									

Remedy Summary—Group Summary of Test Performance

Note: Test remedies are specified in the *Answer Key*. All Mastery Tests are 100 points so the student's score is also the total percentage.

Name	Test 7 Check parts not passed 1 2 3 4 5 6	Total %	Test 8 Check parts not passed 1 2 3 4 5 6 7	Total %	Test 9 Check parts not passed 1 2 3 4 5 7	Total %
1.						
2.						
3.						
4.						
5.						
6.						
7.						
8.						
9.						
10.						
11.						
12.						
13.						
14.						
15.						
16.						
17.						
18.						
19.						
20.						
21.						
22.						
23.						
24.						
25.						
26.						
27.						
28.						
29.						
30.						

Number of students Not Passed = NP

Total number of students = T

Remedy needed if NP/T = 25% or more

Remedy Summary—Group Summary of Test Performance

Note: Test remedies are specified in the *Answer Key*. All Mastery Tests are 100 points so the student's score is also the total percentage.

Name	Test 10 Check parts not passed						Test 10 Total %	Test 11 Check parts not passed							Test 11 Total %	Test 12 Check parts not passed										Test 12 Total %
	1	2	3	4	5	6		1	2	3	4	5	6	7		1	2	3	4	5	6	7	8	9	10	
1.																										
2.																										
3.																										
4.																										
5.																										
6.																										
7.																										
8.																										
9.																										
10.																										
11.																										
12.																										
13.																										
14.																										
15.																										
16.																										
17.																										
18.																										
19.																										
20.																										
21.																										
22.																										
23.																										
24.																										
25.																										
26.																										
27.																										
28.																										
29.																										
30.																										

Number of students Not Passed = NP

Total number of students = T

Remedy needed if NP/T = 25% or more

Connecting Math Concepts

Appendix D

Sample Lessons

- Lesson 16 Presentation Book and Workbook
- Lesson 73 Presentation Book and Workbook

Lesson 16

EXERCISE 1: PLACE-VALUE ADDITION REMEDY

a. I'll say the equation for 70 plus 4. 70 plus 4 equals 74.
- Say the equation for 70 plus 4. (Signal.) *70 plus 4 equals 74.*
(Repeat step a until firm.)

b. Say the equation for 70 plus 9. (Signal.) *70 plus 9 equals 79.*
- Say the equation for 40 plus 9. (Signal.) *40 plus 9 equals 49.*
- (Repeat for 40 + 1, 90 + 6, 30 + 5, 50 + 3.)
(Repeat step b until firm.)

▬ INDIVIDUAL TURNS ▬

Now I'll call on individuals.
(Call on individual students to perform one or two of the following tasks.)

- Say the equation for 40 plus 1. (Call on a student.) *40 + 1 = 41.*
- Say the equation for 90 plus 6. (Call on a student.) *90 + 6 = 96.*
- Say the equation for 30 plus 5. (Call on a student.) *30 + 5 = 35.*

EXERCISE 2: MIXED COUNTING REMEDY

a. My turn to start with 7 and count by ones to 17. I'll get 7 going. Then I'll count.
Listen: Sevennn, 8, 9, 10, 11, 12, 13, 14, 15, 16, 17. Your turn to start with 7 and count to 17.
- What number do you start with? (Signal.) *7.*
- Get 7 going. *Sevennn.* Count. (Tap 10.) *8, 9, 10, 11, 12, 13, 14, 15, 16, 17.*
(Repeat until firm.)

b. Now you'll start with 24 and count to 34. What number will you start with? (Signal.) *24.*
- Get 24 going. *Twenty-fouuur.* Count. (Tap 10.) *25, 26, 27, 28, 29, 30, 31, 32, 33, 34.*
(Repeat step b until firm.)

c. My turn to start with 10 and count backward: Tennn, 9, 8, 7, 6, 5, 4, 3, 2, 1.
- Your turn: Start with 10 and count backward. Get 10 going. *Tennn.* Count backward. (Tap 9.) *9, 8, 7, 6, 5, 4, 3, 2, 1.*
(Repeat step c until firm.)

d. My turn to count by tens to 100: 10, 20, 30, 40, 50, 60, 70, 80, 90, 100.
- Your turn: Count by tens to 100. Get ready. (Tap 10.) *10, 20, 30, 40, 50, 60, 70, 80, 90, 100.*
- Now you'll start with 40 and count by tens to 90. Get 40 going. *Fortyyy.* Count by tens. (Tap 5.) *50, 60, 70, 80, 90.*
- Now you'll start with 20 and count by tens to 90. Get 20 going. *Twentyyy.* Count by tens. (Tap 7.) *30, 40, 50, 60, 70, 80, 90.*
(Repeat step d until firm.)

e. My turn to start with 23 and plus tens to 93: 23, 33, 43, 53, 63, 73, 83, 93.
- Your turn: Start with 23 and plus tens to 93. Get 23 going. *Twenty-threee.* Count. (Tap 7.) *33, 43, 53, 63, 73, 83, 93.*
(Repeat step e until firm.)

f. My turn to start with 28 and plus tens to 98: 28, 38, 48, 58, 68, 78, 88, 98.
- Your turn: Start with 28 and plus tens to 98. Get 28 going. *Twenty-eieieight.* Count. (Tap 7.) *38, 48, 58, 68, 78, 88, 98.*
(Repeat step f until firm.)

EXERCISE 3: TURN-AROUNDS
COMMUTATIVE PROPERTY OF ADDITION

a. (Display:)   [16:3A]

$$5 + 1 = 6$$
$$1 + 7 = 8$$
$$9 + 0 = 9$$

Here's a fact.
- (Point to **5 + 1 = 6.**) Everybody, read the fact. (Touch.) *5 + 1 = 6.*
- (Point to **1.**) My turn to read the turn-around fact. (Touch.) 1 plus 5 equals 6.

b. (Point to **1.**) Your turn to read the turn-around fact. (Touch.) *1 + 5 = 6.*
(Repeat step b until firm.)

c. My turn to say the fact that starts with 5: 5 plus 1 equals 6.
My turn to say the turn-around fact: 1 plus 5 equals 6.

d. Your turn: Say the fact that starts with 5. (Signal.) *5 + 1 = 6.*
• Say the turn-around fact that starts with 1. (Signal.) *1 + 5 = 6.*
(Repeat step d until firm.)

e. (Point to **1 + 7 = 8.**) Here's a new fact. Everybody, read the fact. (Touch.) *1 + 7 = 8.*
• (Point to **7.**) Read the turn-around fact. (Touch.) *7 + 1 = 8.*

f. My turn to say the fact that starts with 1: 1 plus 7 equals 8.
• Your turn: Say the fact that starts with 1. (Signal.) *1 + 7 = 8.*
• Start with 7 and say the turn-around fact. (Signal.) *7 + 1 = 8.*
(Repeat step f until firm.)

g. (Point to **9 + 0 = 9.**) Here's a new fact. Read the fact. (Touch.) *9 + 0 = 9.*
• (Point to **0.**) Read the turn-around fact. (Touch.) *0 + 9 = 9.*

h. Say the fact that starts with 9. (Signal.) *9 + 0 = 9.*
• Start with zero and say the turn-around fact. (Signal.) *0 + 9 = 9.*
(Repeat step h until firm.)

EXERCISE 4: READING 3-DIGIT NUMBERS [REMEDY]

Note: Hundreds numbers like 132 are read without the word **and**: "One hundred thirty-two," not: "A hundred and thirty-two."

a. (Display:)   [16:4A]

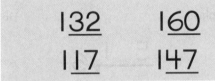

These are hundreds numbers. Part of each hundreds number is underlined. You'll read the underlined part. Then you'll read the whole number.
• (Point to **132.**) Read the underlined part. (Touch.) *32.*
• The whole number is one hundred 32. What's the whole number? (Touch.) *132.*

b. (Point to **117.**) Read the underlined part. (Touch.) *17.*
• What's the whole number? (Touch.) *117.*
c. (Point to **160.**) Read the underlined part. (Touch.) *60.*
• What's the whole number? (Touch.) *160.*
d. (Point to **147.**) Read the underlined part. (Touch.) *47.*
• What's the whole number? (Touch.) *147.*
(Repeat steps b through d until firm.)

■■■■■■ INDIVIDUAL TURNS ■■■■■■

(Call on individual students to identify one or two numbers.)

EXERCISE 5: FACTS
MINUS 1

Note: Present steps a through c for students who enter the program on Lesson 16. Skip steps a through c for students who started on Lesson 1.

a. Listen: When you take away, you **minus.**
• Say **minus.** (Signal.) *Minus.*
(Observe children and give feedback.)
• What's another word for take away? (Signal.) *Minus.*
(Display:)   [16:5A]

$$5 - 1 = 4$$
$$5 - 5 = 0$$

b. (Point to **5 – 1 = 4.**) This says (Touch symbols.) 5 take away 1 equals 4.
My turn to read the equation with another word for take away. (Touch symbols.) 5 minus 1 equals 4.
• Your turn: Read the equation with the word **minus.** Get ready. (Touch symbols.) *5 minus 1 equals 4.*
(Repeat until firm.)
c. (Point to **5 – 5 = 0.**) This equation has a minus sign.
• Read this equation. Get ready. (Touch.) *5 minus 5 equals zero.*
• (Point to **5 – 1.**) Read this equation. (Touch.) *5 minus 1 equals 4.*
(Repeat step c until firm.)
Remember, another word for take away is **minus.**

d. You're going to say facts that minus 1.
- Listen: 9 minus 1 equals 8.
- Say the fact. (Signal.) *9 minus 1 equals 8.*
- Say the fact for 8 minus 1. (Signal.) *8 minus 1 equals 7.*
- Say the fact for 4 minus 1. (Signal.) *4 minus 1 equals 3.*
- Say the fact for 7 minus 1. (Signal.) *7 minus 1 equals 6.*
- Say the fact for 2 minus 1. (Signal.) *2 minus 1 equals 1.*
- Say the fact for 9 minus 1. (Signal.) *9 minus 1 equals 8.*
(Repeat step d until firm.)

EXERCISE 6: ONE MORE THAN

a. I'll tell you a number and you'll tell me the next number when you count by ones.
- Listen: 6. What's the next number? (Signal.) *7.*
(To correct:)
  - Listen: 4, 5, siiiix. What's the next number? (Signal.) *7.*
- Listen: 7. What's the next number? (Signal.) *8.*
- Listen: 9. What's the next number? (Signal.) *10.*
- Listen: 4. What's the next number? (Signal.) *5.*
- Listen: 2. What's the next number? (Signal.) *3.*
(Repeat step a until firm.)

b. The **next** number is **one more.**
- Your turn: 6. What's 1 more than **6?** (Signal.) *7.* So what's 1 plus 6? (Signal.) *7.*
- What's 1 more than **7?** (Signal.) *8.* So what's 1 plus 7? (Signal.) *8.*
- What's 1 more than **8?** (Signal.) *9.* So what's 1 plus 8? (Signal.) *9.*
- What's 1 more than **9?** (Signal.) *10.* So what's 1 plus 9? (Signal.) *10.*

c. Let's do some more.
- Listen: 2. What's 1 more than 2? (Signal.) *3.* So what's 1 plus 2? (Signal.) *3.*

d. (Repeat the following tasks for 5, 3, 4:)
- What's one more than __?
- So what's 1 plus __?
(Repeat steps that were not firm.)

EXERCISE 7: MIXED COUNTING
COUNTING BACKWARD

a. Now we'll count backward. I'll start with 7 and count backward. Sevennn, 6, 5, 4, 3, 2, 1.
- Your turn: Start with 7 and count backward. Get 7 going. *Sevennnn.* Count backward. (Tap 6.) *6, 5, 4, 3, 2, 1.* (Repeat until firm.)

b. I'll start with 37 and count backward to 31.
- What am I going to start with? (Signal.) *37.* Thirty-sevennn, 36, 35, 34, 33, 32, 31.
- Your turn: Start with 37 and count backward to 31. Get 37 going. *Thirty-sevennn.* Count backward. (Tap 6.) *36, 35, 34, 33, 32, 31.* (Repeat step b until firm.)

c. Your turn: Start with 87 and count backward to 81.
- Get 87 going. *Eighty-sevennn.* Count backward. (Tap 6.) *86, 85, 84, 83, 82, 81.* (Repeat step c until firm.)

d. Your turn: Start with 27 and count backward to 21.
- Get 27 going. *Twenty-sevennn.* Count backward. (Tap 6.) *26, 25, 24, 23, 22, 21.* (Repeat step d until firm.)

e. Your turn to count by tens.
- Count by tens to 100. Get ready. (Tap 10.) *10, 20, 30, 40, 50, 60, 70, 80, 90, 100.* (Repeat step e until firm.)

EXERCISE 8: NUMBER FAMILIES

a. (Display:) [16:8A]

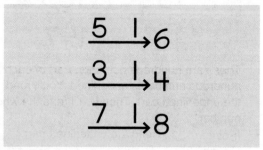

These are **number families.**
- What are they? (Signal.) *Number families.* If you learn number families, you don't have to count to figure out the answer to problems that plus or minus.
Remember: If you know the three numbers in a family, you don't have to count to work plus or minus problems with these numbers.

b. Each number family is on an arrow. Each number family has three numbers.
- How many numbers does each number family have? (Signal.) *Three.*
- (Point to $5 \xrightarrow{\ 1\ } 6$.) How many numbers are in this family? (Touch.) *Three.*
- (Point to $3 \xrightarrow{\ 1\ } 4$.) How many numbers are in this family? (Touch.) *Three.*
- (Point to $7 \xrightarrow{\ 1\ } 8$.) How many numbers are in this family? (Touch.) *Three.*

c. Two numbers are small numbers.
- How many are small numbers? (Signal.) *Two.*
- (Point to **5 and 1**.) The small numbers in this family are 5 and 1. What are the small numbers in this family? (Touch.) *5 and 1.*
- (Point to **6**.) The big number is 6. What's the big number? (Touch.) *6.*

d. (Point to $5 \xrightarrow{\ 1\ } 6$.) Again, what are the small numbers? (Signal.) *5 and 1.*
- What's the big number? (Signal.) *6.*
(Repeat step d until firm.)

e. (Point to **3 and 1**.) The small numbers in this family are 3 and 1. What are the small numbers? (Touch.) *3 and 1.*
- (Point to **4**.) This is the big number. What's the big number? (Touch.) *4.*

f. (Point to $3 \xrightarrow{\ 1\ } 4$.) What are the small numbers? (Signal.) *3 and 1.*
What's the big number? (Signal.) *4.*
- (Point to $7 \xrightarrow{\ 1\ } 8$.) What are the small numbers in this family? (Touch.) *7 and 1.*
What's the big number? (Touch.) *8.*

g. Let's do those again.
- (Point to $5 \xrightarrow{\ 1\ } 6$.) What are the small numbers in this family? (Signal.) *5 and 1.*
What's the big number? (Signal.) *6.*
- (Point to $3 \xrightarrow{\ 1\ } 4$.) What are the small numbers in this family? (Signal.) *3 and 1.*
What's the big number? (Signal.) *4.*
- (Point to $7 \xrightarrow{\ 1\ } 8$.) What are the small numbers in this family? (Signal.) *7 and 1.*
What's the big number? (Signal.) *8.*
(Repeat step g until firm.)

EXERCISE 9: TURN-AROUNDS
COMMUTATIVE PROPERTY OF ADDITION

a. Listen: 4 plus 1.
Here's the turn-around: 1 plus 4.
- Say 4 plus 1. (Signal.) *4 plus 1.*
- Say the turn-around. (Signal.) *1 plus 4.*

b. Listen: 8 plus 3.
Here's the turn-around: 3 plus 8.
- Say 8 plus 3. (Signal.) *8 plus 3.*
- Say the turn-around. (Signal.) *3 plus 8.*

c. Say 1 plus 6. (Signal.) *1 plus 6.*
- Say the turn-around. (Signal.) *6 plus 1.*
(Repeat steps b and c until firm.)

d. Listen: 4 plus 1 equals 5. Say the fact. (Signal.) *4 + 1 = 5.*
If 4 plus 1 equals 5, 1 plus 4 equals 5.
- Say the fact for 4 plus 1. (Signal.) *4 + 1 = 5.*
- Start with 1 and say the turn-around fact. (Signal.) *1 + 4 = 5.*
(Repeat step d until firm.)

e. Say the fact for 7 plus 1. (Signal.) *7 + 1 = 8.*
Start with 1 and say the turn-around fact. (Signal.) *1 + 7 = 8.*
- Say the fact for 3 plus 1. (Signal.) *3 + 1 = 4.*
Start with 1 and say the turn-around fact. (Signal.) *1 + 3 = 4.*
(Repeat step e until firm.)

EXERCISE 10: WORD PROBLEMS
PLUS/MINUS DISCRIMINATION

a. (Distribute unopened workbooks to students. Hold up workbook opened to Lesson 16.) (Teacher reference:)

This is Lesson 16 of your workbook.
(Point to the number 16.) This number (touch 16) shows that this is the worksheet for Lesson 16.
- Open your workbook to Lesson 16.
(Observe children and give feedback.)
- Find part 1 on your worksheet. ✔
There's a **plus** sign and a **minus** sign next to A.
- Touch the plus sign. ✔
- Touch the minus sign. ✔

b. I'm going to say sentences. You're going to circle the plus sign if the sentence pluses. You're going to circle the minus sign if it minuses.
Here's the sentence for A: Jim took 3 frying pans from the kitchen.

- Did the **kitchen** end up with more frying pans or less frying pans? (Signal.) *Less (frying pans).*
- So did the **kitchen** plus or minus frying pans? (Signal.) *Minus (frying pans).*
- Circle the minus sign for A.
  (Observe children and give feedback.)

c. Here's the sentence for B: 12 dogs went out of the park.

- Did the **park** end up with more dogs or less dogs? (Signal.) *Less (dogs).*
- So did the **park** plus or minus dogs? (Signal.) *Minus (dogs).*
- Circle the minus sign for B.
  (Observe children and give feedback.)

d. Here's the sentence for C: 45 people got on a plane.

- Did the **plane** end up with more people or less people? (Signal.) *More (people).*
- So did the **plane** plus or minus people? (Signal.) *Plus (people).*
- Circle the plus sign for C.
  (Observe children and give feedback.)

e. Here's the sentence for D: Ms. Anderson put 5 tools in the garage.

- Think to yourself if the **garage** ended up with more tools or less tools. ✔
- Did the **garage** plus or minus tools? (Signal.) *Plus (tools).*
- Circle the plus sign for D.
  (Observe children and give feedback.)

f. Here's the sentence for E: 9 children got out of the swimming pool.

- Think to yourself if the **swimming pool** ended up with more children or less children. ✔
- Did the **swimming pool** plus or minus children? (Signal.) *Minus (children).*
- Circle the minus sign for E.
  (Observe children and give feedback.)

g. Here's the sentence for F: Millie put 9 cookies on the table.

- Think to yourself if the **table** ended up with more cookies or less cookies. ✔

- Did the **table** plus or minus cookies? (Signal.) *Plus (cookies).*
- Circle the plus sign for F.
  (Observe children and give feedback.)

EXERCISE 11: WRITING TWO-PART NUMBERS

a. Find part 2 on your worksheet. ✔
(Teacher reference:)

a. ___   d. ___
b. ___   e. ___
c. ___   f. ___

You're going to write numbers that have two parts.

- Touch space A. ✔
- You'll write 31 in space A. What number? (Signal.) *31.*
- Write 31.
  (Observe children and give feedback.)
  (Display:) W   [16:11A]

> **a.** 3l

Here's what you should have.

- (Point to **31**.) What number? (Touch.) *31.*

b. Touch space B. ✔
- You'll write 18 in space B. What number? (Signal.) *18.*
- Write 18.
  (Observe children and give feedback.)
  (Add to show:)   [16:11B]

> **b.** l8

Here's what you should have.

- (Point to **18**.) What number? (Touch.) *18.*

c. (Repeat the following tasks for C, 12; D, 53; E, 71; F, 98:)
- Touch space ___.
- You'll write ___ in space ___. What number?
- Write ___.
  (Observe children and give feedback.)
  (Add to show:)   [16:11C-F]
  Here's what you should have.
- (Point to ___.) What number?
  (Teacher reference:)

a. 3l	**d.** 53
b. l8	**e.** 7l
c. l2	**f.** 98

EXERCISE 12: NUMBER LINE
TEEN NUMBERS

a. Find part 3 on your worksheet. ✔
(Teacher reference:)

11 12 __ 14 __ __ 17 __ __

This is a number line for teen numbers, but some of the numbers are missing.
- Raise your hand when you know the first missing number. ✔
- What's the first missing number? (Signal.) *13.*
b. Raise your hand when you know the next missing number. ✔
- What's the next missing number? (Signal.) *15.* You'll write the missing numbers later as part of your independent work.

EXERCISE 13: INDEPENDENT WORK

a. Find part 4 on your worksheet. ✔
(Teacher reference:)

Part 4		
a. $5 + 1 =$	d. $3 + 1 =$	g. $6 + 1 =$
b. $9 + 1 =$	e. $4 + 1 =$	h. $8 + 1 =$
c. $2 + 1 =$	f. $7 + 1 =$	i. $1 + 1 =$

Part 5		
a.	b.	c.
$4 + 0 =$	$2 + 0 =$	$7 + 0 =$
$4 + 1 =$	$2 + 1 =$	$7 + 1 =$
$4 + 2 =$	$2 + 2 =$	$7 + 2 =$

Part 6		
a. $30 + 10 =$	b. $50 + 10 =$	c. $80 + 10 =$

These problems plus 1. Later, you'll write answers to the problems.
b. Find part 5 on your worksheet. ✔
Each item has a plus zero, a plus-1, and a plus-2 problem. You'll write the answers to each problem.
c. Find part 6 on your worksheet. ✔
These problems plus 10. Later you'll complete each equation.
d. Complete the worksheet. Finish writing the numbers for the number line. Then work all of the problems in part 4, part 5, and part 6. **(Observe children and mark incorrect responses on children's worksheets as you give feedback.)**

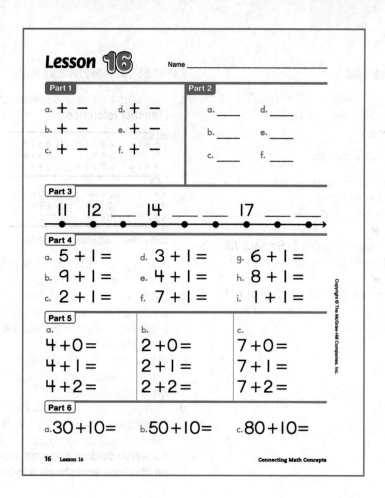

Lesson 16

Name _____

Part 1

a. + −   d. + −

b. + −   e. + −

c. + −   f. + −

Part 2

a. ____   d. ____

b. ____   e. ____

c. ____   f. ____

Part 3

11 12 ___ 14 ___ ___ 17 ___ ___

Part 4

a. 5 + 1 =   d. 3 + 1 =   g. 6 + 1 =

b. 9 + 1 =   e. 4 + 1 =   h. 8 + 1 =

c. 2 + 1 =   f. 7 + 1 =   i. 1 + 1 =

Part 5

a.
4 + 0 =
4 + 1 =
4 + 2 =

b.
2 + 0 =
2 + 1 =
2 + 2 =

c.
7 + 0 =
7 + 1 =
7 + 2 =

Part 6

a. 30 + 10 =   b. 50 + 10 =   c. 80 + 10 =

16 Lesson 16   Connecting Math Concepts

Lesson

EXERCISE 1: NUMBER FAMILIES
MISSING NUMBER IN FAMILY

a. (Display:) [73:1A]

$$\begin{array}{ccc}
\underset{6}{\longrightarrow}11 & \underset{7}{\longrightarrow}9 & \underset{6}{\longrightarrow}12 \\[6pt]
\underset{2}{\longrightarrow}5 & \underset{6\ \ 4}{\longrightarrow}\_ & \underset{10\ \ 6}{\longrightarrow}\_ \\[6pt]
\underset{6\ \ 1}{\longrightarrow}\_ & \underset{2}{\longrightarrow}10 & \underset{3}{\longrightarrow}9
\end{array}$$

You're going to say the problem for the missing number in each family.

b. (Point to $\underset{6}{\longrightarrow}$11.) Say the problem for the missing number. Get ready. (Touch.) *11 minus 6.*
- What's 11 minus 6? (Signal.) *5.*

c. (Point to $\underset{2}{\longrightarrow}$5.) Say the problem for the missing number. (Touch.) *5 minus 2.*
- What's 5 minus 2? (Signal.) *3.*

d. (Repeat the following tasks for remaining families:)

(Point to __.)	Say the problem for the missing number.	What's __?	
$\underset{6\ \ 1}{\longrightarrow}$_	*6 + 1*	6 + 1	7
$\underset{7}{\longrightarrow}$9	*9 − 7*	9 − 7	2
$\underset{6\ \ 4}{\longrightarrow}$_	*6 + 4*	6 + 4	10
$\underset{2}{\longrightarrow}$10	*10 − 2*	10 − 2	8
$\underset{6}{\longrightarrow}$12	*12 − 6*	12 − 6	6
$\underset{10\ \ 6}{\longrightarrow}$_	*10 + 6*	10 + 6	16
$\underset{3}{\longrightarrow}$9	*9 − 3*	9 − 3	6

(Repeat for families that were not firm.)

EXERCISE 2: 3-D OBJECTS

a. (Display:) [73:2A]

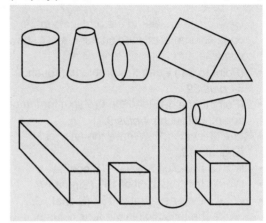

Last time you learned about cylinders. A cylinder has a circle on each end and the circles are the same size.

- (Point to ⬭.) Does this object have a circle on each end? (Touch.) *Yes.*
- Are the circles the same size? (Touch.) *Yes.*
- So is this a cylinder? (Touch.) *Yes.*

b. (Point to ⬭.) Does this object have a circle on each end? (Touch.) *Yes.*
- Are the circles the same size? (Touch.) *No.*
- So is this a cylinder? (Touch.) *No.*

c. Some of these objects are cylinders. Some are cubes.
- (Point to ⬭.) Is this a cylinder? (Touch.) *Yes.*

d. (Point to △.) Is this a cylinder? (Touch.) *No.*
- Is this a cube? (Touch.) *No.*

e. (Point to ⬭.) Is this a cylinder? (Touch.) *No.*
- Is this a cube? (Touch.) *No.*
- Why not? (Call on a student. Idea: *The faces are not all squares.*)

f. (Point to ⬭.) Is this a cylinder? (Touch.) *No.*
- Is this a cube? (Touch.) *Yes.*

g. (Point to ⎮.) Is this a cylinder? (Touch.) *Yes.*

h. (Point to ⬭.) Is this a cylinder? (Touch.) *No.*
- Why not? (Call on a student. Idea: *The circles are not the same size.*)

i. (Point to ⬭.) Is this a cylinder? (Touch.) *No.*
- What is it? (Touch.) *(A) cube.*

EXERCISE 3: COLUMN PROBLEMS
CARRYING PRESKILL

a. (Display:) W   [73:3A]

```
  5 4
+ 3 9
```

Listen: The answer to the problems in the ones column is a two-digit number. I'll show you where to write the tens digit.

- (Point to **54**.) Read this problem. (Touch.) *54 plus 39.*
- (Point to the **4**.) Read the problem for the ones column. (Touch.) *4 plus 9.*
- The answer is 13. What's the answer? (Signal.) *13.*
- Is 13 a two-digit answer? (Signal.) *Yes.*
- What's the tens digit of 13? (Signal.) *1.*
- What's the ones digit of 13? (Signal.) *3.*
  I write the tens digit in the tens column, and the 3 in the ones column.
  Watch:
  (Add to show:)   [73:3B]

```
   1
  5 4
+ 3 9
    3
```

b. (Display:) W   [73:3C]

```
  2 9
+ 4 1
```

- (Point to **29**.) Read this problem. (Touch.) *29 plus 41.*
- (Point to the **9**.) Read the problem for the ones column. (Touch.) *9 plus 1.*
- What's the answer? (Signal.) *10.*
- Is 10 a two-digit answer? (Signal.) *Yes.*
- What's the tens digit of 10? (Signal.) *1.*
  I write that digit in the tens column and the zero in the ones column.
  Watch:
  (Add to show:)   [73:3D]

```
   1
  2 9
+ 4 1
    0
```

c. (Display:) W   [73:3E]

```
  1 2
+ 6 8
```

- (Point to **12**.) Read this problem. (Touch.) *12 plus 68.*
- (Point to the **2**.) Read the problem for the ones column. (Touch.) *2 plus 8.*
- What's the answer? (Signal.) *10.*
- Is 10 a two-digit answer? (Signal.) *Yes.*
- What's the tens digit of 10? (Signal.) *1.*
- Where do I write the tens digit? (Signal.) *In the tens column.*
- Where do I write zero? (Signal.) *In the ones column.*
  (Repeat until firm.)
  (Add to show:)   [73:3F]

```
   1
  1 2
+ 6 8
    0
```

d. (Display:) W   [73:3G]

```
  3 6
+ 1 6
```

- (Point to **36**.) Read this problem. (Touch.) *36 plus 16.*
- (Point to the **6**.) Read the problem for the ones column. (Touch.) *6 plus 6.*
- What's the answer? (Signal.) *12.*
- Is 12 a two-digit answer? (Signal.) *Yes.*
- What's the tens digit of 12? (Signal.) *1.*
- Where do I write the tens digit? (Signal.) *In the tens column.*
- Where do I write 2? (Signal.) *In the ones column.*
  (Repeat until firm.)
  (Add to show:)   [73:3H]

```
   1
  3 6
+ 1 6
    2
```

Remember how to write two-digit answers for the ones column.

EXERCISE 4: FACTS

SUBTRACTION

REMEDY

a. (Display:)   [73:4A]

12 – 6	9 – 2	8 – 6
14 – 4	8 – 3	9 – 3
11 – 5	10 – 4	11 – 2

You're going to say the facts for all of these minus problems.

- (Point to **12 – 6.**) Read the problem. Get ready. (Touch.) *12 minus 6.*
- What's 12 minus 6? (Signal.) *6.*
- Say the fact. (Signal.) *12 – 6 = 6.*

b. (Repeat the following tasks for the remaining problems:)
- (Point to __.) Read the problem.
- What's __?
- Say the fact.
(Repeat problems that were not firm.)

EXERCISE 5: MONEY

WRITING DOLLAR AMOUNTS

a. (Display:)   [73:5A]

$3.18	$9.70
$5.98	$15.11
$11.40	

You'll read these dollar amounts.

- (Point to **$3.18.**) Read this. (Signal.) *3 dollars and 18 cents.*
- (Repeat for remaining amounts.)
(Repeat amounts that were not firm.)

b. (Distribute unopened workbooks to students.)
- Open your workbook to Lesson 73 and find part 1.
(Observe children and give feedback.)
(Teacher reference:)

a. $_____   b. $_____   c. $_____

You're going to write dollar amounts. The dollar sign is already written for each amount.

- Touch space A. ✔
- You'll write 6 dollars and 31 cents in space A. What will you write? (Signal.) *6 dollars and 31 cents.*
- Touch the dollar sign for A. ✔
- What do you write after the dollar sign for 6 dollars and 31 cents? (Signal.) *6.*
- What do you write next? (Signal.) *(The) dot.*
- What do you write after the dot? (Signal.) *31.*
- Write 6 dollars and 31 cents.
(Observe children and give feedback.)

c. Check your work.
- Touch the dollar sign for A. ✔
- Touch the number 6. ✔
- Touch the dot. ✔
- Touch the number 31. ✔
- Did you do everything right?

d. Touch the space for B. ✔
You'll write 8 dollars and 70 cents for space B.
- What will you write? (Signal.) *8 dollars and 70 cents.*
- What do you write after the dollar sign for 8 dollars and 70 cents? (Signal.) *8.*
- What do you write next for 8 dollars and 70 cents? (Signal.) *(The) dot.*
- What do you write after the dot? (Signal.) *70.*
- Write 8 dollars and 70 cents.
(Observe children and give feedback.)

e. Check your work.
- Touch the dollar sign for B. ✔
- What's the first thing you wrote? (Signal.) *8.*
- What's the next thing you wrote? (Signal.) *(A) dot.*
- What's the next thing you wrote? (Signal.) *70.*
- Read the amount for B. Get ready. (Signal.) *8 dollars and 70 cents.*

f. Touch the space for C. ✔
You'll write 13 dollars and 59 cents for space C.
- What will you write? (Signal.) *13 dollars and 59 cents.*
- What do you write after the dollar sign for 13 dollars and 59 cents? (Signal.) *13.*
- What do you write next for 13 dollars and 59 cents. (Signal.) *(The) dot.*
- What do you write after the dot? (Signal.) *59.*
- Write 13 dollars and 59 cents.
(Observe children and give feedback.)

g. Check your work.
- Touch the dollar sign for C. ✔
- What's the first thing you wrote? (Signal.) *13.*
- What's the next thing you wrote? (Signal.) *(A) dot.*
- What's the next thing you wrote? (Signal.) *59.*
- Read the amount for C. Get ready. (Signal.) *13 dollars and 59 cents.*

EXERCISE 6: RULER
COUNT BACKWARD

a. Find part 2 on worksheet 73. ✔
(Teacher reference:)

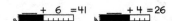

We're going to figure out how many centimeters are shaded for each ruler.
- Touch ruler A. ✔
- Touch the number above the unshaded part. ✔
- How many centimeters is the unshaded part? (Signal.) *6.*
- Touch the number after equals. ✔
- How many centimeters is ruler A? (Signal.) *41.*
- Touch the end of the ruler. ✔
I'll get 41 going and count backward. You'll touch the correct lines. Tell me to stop when you're touching the end of the shaded part.
(Children should be touching the end of the ruler.) Forty-wuuun. (Children touch lines.) 40, 39, 38, 37, 36, 35. *Stop.*
(Repeat until firm.)
b. Your turn to count backward.
- What number will you get going? (Signal.) *41.*
- Touch the end of the ruler. ✔
- Get it going. *Forty-wuuun.* Touch and count. (Tap.) *40, 39, 38, 37, 36, 35, stop.*
(Repeat step b until firm.)
c. How many centimeters is the shaded part? (Signal.) *35.*
- Write 35. ✔
- Touch and read the equation. Get ready. (Signal.) *35 + 6 = 41.*
d. Touch ruler B. ✔
- Touch the number above the unshaded part. ✔
- How many centimeters is the unshaded part? (Signal.) *4.*
- Touch the number after the equals. ✔
- How many centimeters are both parts? (Signal.) *26.*

- You're going to get 26 going and count backward. You'll touch the correct lines and say stop when you get to the end of the shaded part.
- Touch the end of ruler B. ✔
- Get it going. *Twenty-siiix.* Touch and count. (Tap.) *25, 24, 23, 22, stop.*
(Repeat until firm.)
e. How many centimeters is the shaded part? (Signal.) *22.*
- Write 22. ✔
- Touch and read the equation. Get ready. (Signal.) *22 + 4 = 26.*

EXERCISE 7: 3 ADDENDS IN COLUMNS

a. Find part 3 on worksheet 73. ✔
(Teacher reference:)

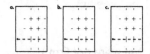

I'll tell you problems that plus three numbers. You're going to write those problems in columns. Only part of the column and row lines are shown and the equals bar is dotted. You'll write the digits and the signs in the right place and make the equals bar.
- Touch where you'll write column problem A. ✔
Listen to problem A: 12 plus 630 plus 7.
- Listen again: 12 plus 630 plus 7. Say problem A. (Signal.) *12 plus 630 plus 7.*
(Repeat until firm.)
b. Put your pencil where you'll write the first digit of 12. ✔
- Put your pencil where you'll write the plus sign. ✔
- Put your pencil where you'll write the first digit of 630. ✔
- Put your pencil where you'll write 7. ✔
- Put your pencil where you'll start the equals bar.
c. Say the problem 12 plus 630 plus 7 for problem A. Get ready. (Signal.) *12 plus 630 plus 7.*
- Write 12 plus 630 plus 7 for problem A. Remember to make the equals bar.
(Observe children and give feedback.)
(Teacher reference:)

d. Touch where you'll write column problem B. ✔
Listen to problem B: 532 plus 64 plus 3.

• Listen again: 532 plus 64 plus 3. Say
problem B. (Signal.) *532 plus 64 plus 3.*
(Repeat until firm.)

• Write 532 plus 64 plus 3 for problem B.
Remember to make the equals bar.
(Observe children and give feedback.)
(Teacher reference:)

e. Touch where you'll write column problem C. ✔
Listen to problem C: 401 plus 2 plus 95.

• Listen again: 401 plus 2 plus 95. Say
problem C. (Signal.) *401 plus 2 plus 95.*
(Repeat until firm.)

• Write 401 plus 2 plus 95 for problem C.
Remember to make the equals bar.
(Observe children and give feedback.)
(Teacher reference:)

Later, you'll work the problems in part 3 as
part of your independent work.

EXERCISE 8: FACTS

ADDITION   REMEDY

a. Find part 4 on worksheet 73. ✔
(Teacher reference:)   R Part H

	₄2 + 7	ₕ6 + 10
ₐ7 + 1	ₑ6 + 6	ᵢ5 + 6
ᵦ4 + 6	ₑ4 + 2	ⱼ6 + 2
ₑ5 + 3	₉3 + 6	

These are plus problems for families you
know. You're going to read each problem and
tell me the answer. Then you'll go back and
work all of the problems.

• Touch and read problem A. Get ready.
(Signal.) *7 plus 1.*

• What's 7 plus 1? (Signal.) *8.*

b. Touch and read problem B. (Signal.) *4 plus 6.*

• What's 4 plus 6? (Signal.) *10.*

c. (Repeat the following tasks for problems C
through J:)

• Touch and read problem __.

• What's __?

d. Complete the equations for all of the problems
in part 4.
(Observe children and give feedback.)

e. Check your work. You'll touch and read
each fact.

• Fact A. (Signal.) *7 + 1 = 8.*

• (Repeat for:) B, *4 + 6 = 10;* C, *5 + 3 = 8;*
D, *2 + 7 = 9;* E, *6 + 6 = 12;* F, *4 + 2 = 6;*
G, *3 + 6 = 9;* H, *6 + 10 = 16;* I, *5 + 6 = 11;*
J, *6 + 2 = 8.*

EXERCISE 9: WORD PROBLEMS (COLUMNS)

a. Find part 5 on worksheet 73. ✔
(Teacher reference:)

You're going to write the symbols for word
problems in columns and work them. Only
part of the column and row lines are shown
and the equals bars are dotted. You'll write the
numbers and the sign in the right places, and
you'll make the equals bar.

b. Touch where you'll write the symbols for
problem A. ✔
Listen to problem A: There were 580 apples
on a tree. Jan and Jerry picked 430 of
those apples. How many apples were still
on the tree?

• Listen again and write the symbols for both
parts: There were 580 apples on a tree.
Jan and Jerry picked 430 of those apples.

• Everybody, touch and read problem A. Get
ready. (Signal.) *580 minus 430.*

c. Work problem A. Put your pencil down when
you know how many apples were still on
the tree.
(Observe children and give feedback.)

• Read the problem and the answer you wrote
for A. Get ready. (Signal.) *580 − 430 = 150.*

• How many apples were still on the tree?
(Signal.) *150.*

d. . Touch where you'll write the symbols for problem B. ✔

Listen to problem B: Mr. Briggs painted 113 pictures last year. He painted 46 this year. How many pictures did he paint altogether?

- Listen again and write the symbols for both parts: Mr. Briggs painted 113 pictures last year. He painted 46 this year.
- Everybody, touch and read problem B. Get ready. **(Signal.)** *113 plus 46.*

e. Work problem B. Put your pencil down when you know how many pictures Mr. Briggs painted altogether.
(Observe children and give feedback.)
(Answer key:)

- Read the problem and the answer you wrote for B. Get ready. **(Signal.)** *113 + 46 = 159.*
- How many pictures did Mr. Briggs paint altogether? **(Signal.)** *159.*

EXERCISE 10: INDEPENDENT WORK

a. Find part 6 on worksheet 73. ✔
(Teacher reference:)

$$\frac{6\quad 6}{\underline{\qquad\qquad}}\!_{,12}$$

There are only two facts for this number family. You'll write the facts in the spaces below.

b. Turn to the other side of worksheet 73 and find part 7. ✔
(Teacher reference:)

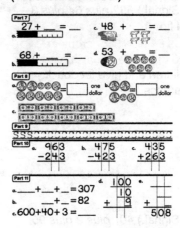

You'll complete the equations for the rulers and the objects.

In part 8, you'll circle the words one dollar or write the cents for each group of coins. You'll write an equals and the number of dollars to show what the group of bills is worth.

In part 9, you'll write the row of Ss.

You'll work the problems in parts 10 and 11.

c. Complete worksheet 73. Remember to work the problems in part 3 and write the facts for part 6 on the other side of it.
(Observe children and mark incorrect responses on children's worksheets as you give feedback.)

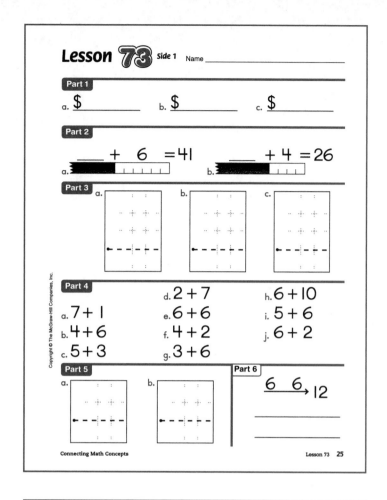

Lesson 73 · Side 1   Name _____

Part 1

a. $ _____   b. $ _____   c. $ _____

Part 2

a. ____ + 6 = 41   b. ____ + 4 = 26

Part 3

a. ☐   b. ☐   c. ☐

Part 4

a. 7 + 1
b. 4 + 6
c. 5 + 3
d. 2 + 7
e. 6 + 6
f. 4 + 2
g. 3 + 6
h. 6 + 10
i. 5 + 6
j. 6 + 2

Part 5

a. ☐   b. ☐

Part 6

$$\underline{6} \quad \underline{6} \rightarrow 12$$

_____

_____

Connecting Math Concepts   Lesson 73   **25**

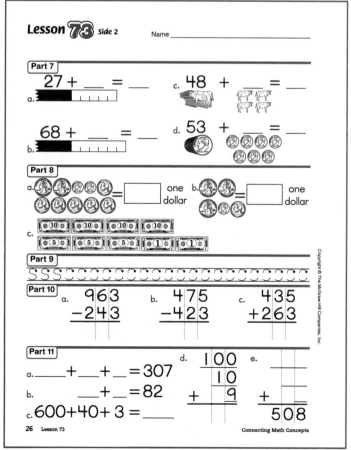

Lesson 73 · Side 2   Name _____

Part 7

a. 27 + ____ = ____

b. 68 + ____ = ____

c. 48 + ____ = ____

d. 53 + ____ = ____

Part 8

a. ☐ = ☐ one dollar   b. ☐ = ☐ one dollar

c. ☐ ☐ ☐ ☐
   ☐ ☐ ☐ ☐ ☐

Part 9

$ $ $ ...

Part 10

a. 963
   −243

b. 475
   −423

c. 435
   +263

Part 11

a. ____ + ____ + ____ = 307

b. ____ + ____ = 82

c. 600 + 40 + 3 = ____

d. 100
     10
   + 9
   ─────

e. ____
     ____
   + ____
   ─────
     508

26   Lesson 73   Connecting Math Concepts